ARTHROPODS OF MEDICAL AND VETERINARY IMPORTANCE:

A Checklist of Preferred Names and Allied Terms

ARTHROPODS OF MEDICAL AND VETERINARY IMPORTANCE:

A Checklist of Preferred Names and Allied Terms

Compiled by

A.R. Pittaway

C·A·B International

C·A·B International
Wallingford
Oxon OX10 8DE
UK

Tel: Wallingford (0491) 32111
Telex: 847964 (COMAGG G)
Telecom Gold/Dialcom: 84: CAU001
Fax: (0491) 33508

A catalogue entry is available from the British Library

ISBN 0 85198 741 9

Printed and bound in the UK

CONTENTS

PREFACE

BACKGROUND

This checklist is intended to provide a single, reliable source for checking the scientific names and taxonomic position of most important species and genera of arthropods in the fields of medical and veterinary entomology, where the term entomology is used in its broadest sense to include all arthropods. It also includes organisms which are used to control such arthropods, intermediate hosts, certain wild natural enemies, organisms carried by them and which cause diseases in man and economically important animals, and all important synonyms. It is also intended to introduce a measure of consistency to the usage of such names; accurate and consistent nomenclature is essential if information is to be successfully communicated to other people. For various reasons, not all authors use the same name for the same organism, be it because they follow different classification schemes, do not have access to recently published taxonomic revisions, do not know of such revisions or lack and reference documents. Additionally, spelling errors are common.

All the scientific names given were derived from CAB ABSTRACTS, an abstracts database maintained by CAB International. Once extracted, they were checked against internationally recognized references or by leading specialists, even though they had been checked prior to their inclusion into the database by reference to a card index maintained by the CAB International Institute of Entomology. This contains a card for each name that has been recorded in the *Review of Applied Entomology, Series B* (now *Review of Medical and Veterinary Entomology*) since its first issue in 1913.

LAYOUT

The organisms are split up into separate sections on arthropods, nematodes, entomogenous fungi, larvivorous fish, microorganisms other than viruses, arthropod–transmitted viruses and any other organisms, such as planarians. The names are arranged alphabetically on the left (whether correct or incorrect), followed by the author of the species name and, where applicable, the correct name of the organism to the right preceded by the term '**see**'. Each full name is preceded by the generic name with the relevant family, order, subclass or class (if known) to the right.

UPDATING

Although the names in the lists have been carefully scrutinized during compilation, a few errors have almost certainly been included. Additionally, taxonomic revisions are constantly being published and it is inevitable that changes will need to be made to any future editions. Any errors or changes found should be brought to the attention of the compiler; these will be rectified on the MEDANI database maintained at CAB International.

ACKNOWLEDGEMENTS

In a work such as this one must by necessity elicit the help of a number of specialists to both check and contribute towards the final product. I would like to thank the following people for their help: Ms Joan Harvey, Mr Alan Wood, Dr Don Macfarlane, Mrs Lesley Bell–Sakyi, Dr Michèle Williams, Dr Ronny Larsson, Dr John Baker, Dr David Hunt, Dr Lynda Gibbons, Dr Harry Evans, Dr L.R. Hill, Dr Patricia Nuttall, Dr Arlene Jones, Dr Duncan Brown, Prof Elizabeth Canning, Prof K. Vickerman and the staff of The Natural History Museum in London.

Tony Pittaway
CAB INTERNATIONAL
Wallingford
OX10 8DE
UK

ACKNOWLEDGEMENTS

ARTHROPODS

The following are the most important arthropods encountered in medical and veterinary entomology. The family and order/another taxon of each is given to the right, except where a synonym is present.

A

Ablabesmyia — Chironomidae, Diptera

Abonnencius — **see** *Phlebotomus*

Acanthochela — Laelapidae, Acari
Acanthochela chilensis Ewing

Acanthochondria — Chondracanthidae, Copepoda
Acanthochondria diastema Kabata

Acanthocyclops — Cyclopidae, Copepoda
Acanthocyclops vernalis (Fischer)
Acanthocyclops viridis (Jurine) — **see** *Megacyclops viridis*

Acantholepis — Formicidae, Hymenoptera

Acanthophthirius — Myobiidae, Acari
Acanthophthirius nycticeius Fain & Whitaker

Acanthoscurria — Theraphosidae, Araneae

Acarus — Acaridae, Acari
Acarus farris (Oudemans)
Acarus griffithsi Ranganath & ChannaBasavanna
Acarus monopsyllus Fain & Schwan
Acarus scabiei Linnaeus — **see** *Sarcoptes scabiei*
Acarus siro Linnaeus

Achaearanea — Theridiidae, Araneae
Achaearanea tabulata Levi
Achaearanea tepidariorum (Koch)

Acheta — Gryllidae, Orthoptera
Acheta domesticus (Linnaeus)

Achtheres — Lernaeopodidae, Copepoda
Achtheres percarum Nordmann

Acotyledon — **see** *Caloglyphus*
Acotyledon neotomae Fain & Whitaker — **see** *Caloglyphus neotomae*
Acotyledon paradoxa Oudemans — **see** *Caloglyphus paradoxus*

Acricotopus — Chironomidae, Diptera
Acricotopus lucens (Zetterstedt)
Acricotopus lucidus (Staeger) — **see** *Acricotopus lucens*

Acritus — Histeridae, Coleoptera
Acritus nigricornis (Hoffmann)

Acrolytta — Meloidae, Coleoptera
Acrolytta neivai (Denier) — **see** *Lytta neivai*
Acrolytta nigropicta (Denier) — **see** *Picnoseus nigropictus*

Acroneuria — Perlidae, Plecoptera
Acroneuria lycorias (Newman)

Acropsylla *Acropsylla traubi* Lewis	Leptopsyllidae, Siphonaptera
Acrotrichis *Acrotrichis africana* Johnson *Acrotrichis palmi* Johnson *Acrotrichis propinqua* Johnson *Acrotrichis sericans* (Heer) *Acrotrichis setigera* Johnson	Ptiliidae, Coleoptera
Acyrthosiphon *Acyrthosiphon pisum* (Harris)	Aphididae, Hemiptera
Adia	Anthomyiidae, Diptera
Adiscochaeta *Adiscochaeta abnormis* Enderlein	Sarcophagidae, Diptera
Adoratopsylla *Adoratopsylla antiquorum* (Rothschild)	Hystrichopsyllidae, Siphonaptera
Aedeomyia *Aedeomyia catasticta* Knab *Aedeomyia squamipennis* (Lynch Arribálzaga)	Culicidae, Diptera

Aedes Culicidae, Diptera
Aedes abserratus (Felt & Young)
Aedes aegypti aegypti (Linnaeus)
Aedes aegypti formosus (Walker)
Aedes africanus (Theobald)
Aedes agrestis Barraud
Aedes albifasciatus (Macquart)
Aedes albineus Séguy **see** *Aedes caspius*
Aedes alboannulatus (Macquart)
Aedes albocephalus (Theobald)
Aedes albodorsalis Fontenille & Brunhes
Aedes albolineatus (Theobald)
Aedes albopictus (Skuse)
Aedes alboscutellatus (Theobald)
Aedes alcasidi Huang
Aedes alternans (Westwood)
Aedes ambreensis Rodhain & Boutonnier
Aedes ananae Knight & Laffoon
Aedes angustivittatus Dyar & Knab
Aedes annulipes (Meigen)
Aedes antuensis Su, Wang & Li **see** *Aedes pingpaensis*
Aedes apicoargenteus (Theobald)
Aedes argenteopunctatus (Theobald)
Aedes atlanticus Dyar & Knab
Aedes atropalpus (Coquillett)
Aedes aurifer (Coquillett)
Aedes australis (Erichson)
Aedes australis (Taylor) **see** *Aedes tremulus*
Aedes bahamensis Berlin
Aedes baisasi Knight & Hull
Aedes bambiotai Geoffroy
Aedes bancoi Geoffroy
Aedes bancroftianus Edwards
Aedes bekkui Mogi
Aedes beklemishevi Denisova **see** *Aedes euedes*
Aedes berlandi Séguy
Aedes berlini Schick
Aedes brelandi Zavortink
Aedes bromeliae (Theobald)

Aedes caballus (Theobald)
Aedes campestris Dyar & Knab
Aedes camptorhynchus (Thomson)
Aedes canadensis (Theobald)
Aedes cantans (Meigen)
Aedes cantator (Coquillett)
Aedes capensis Edwards
Aedes carmenti Edwards
Aedes caspius (Pallas)
Aedes caspius dorsalis (Meigen)   **see** *Aedes dorsalis*
Aedes cataphylla Dyar
Aedes cinereus Meigen
Aedes circumluteolus (Theobald)
Aedes communis (DeGeer)
Aedes cooki Belkin
Aedes coulangesi Rodhain & Boutonnier
Aedes cretatus Delfinado
Aedes culicinus Edwards
Aedes cumminsii cumminsii (Theobald)
Aedes cumminsii form *mediopunctatus* (Theobald)
Aedes cyprioides Danilov & Stupin
Aedes cyprius Ludlow
Aedes cyrtolabis Edwards
Aedes dalzieli (Theobald)
Aedes dentatus (Theobald)
Aedes detritus (Haliday)
Aedes diantaeus Howard, Dyar & Knab
Aedes dorsalis (Meigen)
Aedes duplex Martini
Aedes dupreei (Coquillett)
Aedes durbanensis (Theobald)
Aedes echinus (Edwards)
Aedes epactius Dyar & Knab
Aedes esoensis esoensis Yamada
Aedes esoensis rossicus Dolbeshkin *et al.*
Aedes euedes Howard, Dyar & Knab
Aedes excrucians (Walker)
Aedes fitchii (Felt & Young)
Aedes flavescens (Müller)
Aedes flavipennis (Giles)
Aedes flavopictus flavopictus Yamada
Aedes flavopictus miyarai Tanaka *et al.*
Aedes flavus (Motshulsky)   **see** *Aedes flavescens*
Aedes flavus Yamada   *nomen nudum*
Aedes fluviatilis (Lutz)
Aedes formosensis Yamada
Aedes fowleri (Charmoy)
Aedes fryeri (Theobald)
Aedes fulvus fulvus (Wiedemann)
Aedes fulvus pallens Ross
Aedes furcifer (Edwards)
Aedes galloisi Yamada
Aedes geminus Peus
Aedes geniculatus (Olivier)
Aedes gilliesi van Someren
Aedes grassei Doucet
Aedes grossbecki Dyar & Knab
Aedes guamensis Farner & Bohart
Aedes gubernatoris (Giles)
Aedes haworthi Edwards
Aedes hebrideus Edwards
Aedes hemisurus Dyar & Knab
Aedes hendersoni Cockerell
Aedes hexodontus Dyar

Aedes hirsutus (Theobald)
Aedes idahoensis (Theobald)   **see** *Aedes spencerii idahoensis*
Aedes impiger (Walker)
Aedes implicatus Vockeroth
Aedes increpitus Dyar
Aedes ingrami Edwards
Aedes intermedius Danilov & Gornostaeva
Aedes intrudens Dyar
Aedes irritans (Theobald)
Aedes japonicus (Theobald)
Aedes juppi McIntosh
Aedes kapretwae Edwards
Aedes katherinensis Woodhill   **see** *Aedes scutellaris katherinensis*
Aedes kennethi Muspratt
Aedes kesseli Huang & Hitchcock
Aedes kochi (Dönitz)
Aedes koreicoides Sasa, Kano & Hayashi
Aedes koreicus (Edwards)
Aedes krombeini Huang
Aedes krymmontanus Alekseev
Aedes lambrechti van Someren
Aedes lepidonotus (Edwards)
Aedes leucocelaenus Dyar & Shannon   **see** *Haemagogus leucocelaenus*
Aedes leucomelas (Meigen)
Aedes lilii (Theobald)
Aedes lineatopennis (Ludlow)
Aedes longipalpis (Grünberg)
Aedes lugubris Barraud
Aedes luridus McIntosh
Aedes luteocephalus (Newstead)
Aedes madagascarensis van Someren
Aedes malayensis Colless   **see** *Aedes scutellaris malayensis*
Aedes mariae (Edmond & Etienne Sergent)
Aedes mascarensis MacGregor
Aedes masoalensis Fontenille & Brunhes
Aedes mathioti Fontenille & Brunhes
Aedes mcintoshi Huang
Aedes mediopunctatus (Theobald)   **see** *Aedes cumminsii* form *mediopunctatus*
Aedes mediovittatus (Coquillett)
Aedes melanimon Dyar
Aedes metallicus (Edwards)
Aedes mickevichae Huang
Aedes mikrokopion Knight & Harrison
Aedes mitchellae (Dyar)
Aedes monocellatus Marks
Aedes montchadskyi Dubitskiĭ
Aedes monticola Belkin & McDonald
Aedes moucheti Ravaonjanahary & Brunhes
Aedes natronius Edwards
Aedes neoafricanus Cornet, Valade & Dieng
Aedes nevadensis Chapman & Barr
Aedes nigripes (Zetterstedt)
Aedes nigromaculis (Ludlow)
Aedes niphadopsis Dyar & Knab
Aedes niveus (Ludlow)
Aedes normanensis (Taylor)
Aedes notoscriptus (Skuse)
Aedes novalbopictus Barraud
Aedes opok Corbet & van Someren
Aedes pecuniosus Edwards
Aedes pembaensis Theobald
Aedes periskelatus (Giles)
Aedes phoeniciae Coluzzi & Sabatini
Aedes pingpaensis Chang

Aedes pionips Dyar
Aedes platylepidus Knight & Hull
Aedes poicilius (Theobald)
Aedes polynesiensis Marks
Aedes priestleyi (Taylor)   **see** *Aedes purpureus*
Aedes procax (Skuse)
Aedes provocans (Walker)
Aedes pseudalbopictus (Borel)
Aedes pseudomediofasciatus (Theobald)
Aedes pseudonigeria (Theobald)
Aedes pseudoscutellaris (Theobald)
Aedes pulchritarsis (Rondani)
Aedes pullatus (Coquillett)
Aedes punctodes Dyar
Aedes punctor (Kirby)
Aedes punctothoracis (Theobald)
Aedes purpureipes Aitken
Aedes purpureus (Theobald)
Aedes quasiferinus Mattingley
Aedes quasirusticus Torres Cañamares
Aedes refiki Medschid
Aedes reinerti Rattanarithikul & Harrison
Aedes rempeli Vockeroth
Aedes riparioides Su & Zhang
Aedes riparius Dyar & Knab
Aedes riversi Bohart & Ingram
Aedes rossicus Dolbeshkin *et al.*   **see** *Aedes esoensis rossicus*
Aedes rubrithorax (Macquart)
Aedes rupestris Dobrotworsky
Aedes rusticus (Rossi)
Aedes sagax (Skuse)
Aedes saimedres Huang
Aedes samoanus (Grünberg)
Aedes scapularis (Rondani)
Aedes scutellaris katherinensis Woodhill
Aedes scutellaris malayensis Colless
Aedes scutellaris scutellaris (Walker)
Aedes seatoi Huang
Aedes sergievi Danilov, Markovich & Proskuryakova
Aedes serratus (Theobald)
Aedes sierrensis (Ludlow)
Aedes simpsoni (Theobald)
Aedes sintoni (Barraud)
Aedes sollicitans (Walker)
Aedes spencerii idahoensis (Theobald)
Aedes spencerii spencerii (Theobald)
Aedes squamiger (Coquillett)
Aedes sticticus (Meigen)
Aedes stimulans (Walker)
Aedes stokesi Evans
Aedes subalbirostris Klein & Marks
Aedes subdiversus Martini
Aedes sudanensis (Theobald)
Aedes tabu Ramalingam & Belkin   **see** *Aedes tongae tabu*
Aedes taeniorhynchus (Wiedemann)
Aedes tarsalis (Newstead)
Aedes taylori Edwards
Aedes terrens (Walker)
Aedes thelcter Dyar
Aedes theobaldi (Taylor)
Aedes thibaulti Dyar & Knab
Aedes thomsoni (Theobald)
Aedes tiptoni Grjebine
Aedes togoi (Theobald)

Aedes tongae tabu Ramalingam & Belkin
Aedes tormentor Dyar & Knab
Aedes tremulus (Theobald)
Aedes triseriatus (Say)
Aedes trivittatus (Coquillett)
Aedes unidentatus McIntosh
Aedes unilineatus (Theobald)
Aedes varipalpus (Coquillett)
Aedes versicolor (Barraud)
Aedes vexans nipponii (Theobald)
Aedes vexans vexans (Meigen)
Aedes vigilax (Skuse)
Aedes vittatus (Bigot)
Aedes vittiger (Skuse)
Aedes w-albus (Theobald)
Aedes wasselli Marks
Aedes watteni Lien
Aedes yusafi Barraud
Aedes zammitii (Theobald)
Aedes zoosophus Dyar & Knab

Aedimorphus   **see** *Aedes*

Aega   Aegidae, Isopoda
Aega leptonica Bruce

Aegyptogalumna   Galumnidae, Acari
Aegyptogalumna mastigophora Al-Assiuty *et al.*

Aetheca   Ceratophyllidae, Siphonaptera
Aetheca wagneri (Baker)

Afrocimex   Cimicidae, Hemiptera
Afrocimex leleupi Schouteden

Afrolistrophorus   Listrophoridae, Acari
Afrolistrophorus apodemi Fain
Afrolistrophorus laticoxa Fain & Lukoschus
Afrolistrophorus neacomys Fain & Lukoschus
Afrolistrophorus obesus Fain & Lukoschus
Afrolistrophorus punctatus Fain & Lukoschus

Afropolonia   Trombiculidae, Acari
Afropolonia tgifi Goff

Afrotrombicula   Trombiculidae, Acari

Agastopsylla   Hystrichopsyllidae, Siphonaptera
Agastopsylla boxi Jordan & Rothschild
Agastopsylla nylota euneomys Lewis
Agastopsylla nylota nylota Traub

Agelenopsis   Agalenidae, Araneae
Agelenopsis aperta (Gertsch)

Aglaodiaptomus   Diaptomidae, Copepoda
Aglaodiaptomus leptopus (Forbes)

Agnetina   Perlidae, Plecoptera
Agnetina capitata (Pictet)

Albionella   Lernaeopodidae, Copepoda
Albionella centroscyllii (Hansen)
Albionella fabricii Rubec & Hogans

ARTHROPODS OF MEDICAL AND VETERINARY IMPORTANCE

Aldrichina *Aldrichina grahami* (Aldrich)	Calliphoridae, Diptera
Alella *Alella macrotrachelus* (Brian) *Alella tarakihi* Hewitt & Blackwell	Lernaeopodidae, Copepoda
Aleochara *Aleochara bipustulata* (Linnaeus) *Aleochara curtula* (Goeze) *Aleochara lata* Gravenhorst *Aleochara notula* Erichson *Aleochara tristis* Gravenhorst *Aleochara verna* Say	Staphylinidae, Coleoptera
Aleuroglyphus *Aleuroglyphus ovatus* (Troupeau)	Acaridae, Acari
Alleustathia	Eustathiidae, Acari
Alliphis *Alliphis halleri* (G. & R. Canestrini) *Alliphis necrophilus* Christie	Eviphididae, Acari
Allobrueelia *Allobrueelia abluda* Zlotorzycka *Allobrueelia amsel* Eichler	Philopteridae, Phthiraptera **see** *Brueelia abluda* **see** *Brueelia amsel*
Allocyclops	Cyclopidae, Copepoda
Allodermanyssus *Allodermanyssus sanguineus* (Hirst)	**see** *Liponyssoides* **see** *Liponyssoides sanguineus*
Alloptes *Alloptes stercorarii* Dubinin	Alloptidae, Acari
Alluaudomyia *Alluaudomyia epsteini* Giles & Wirth *Alluaudomyia melanesiae* Clastrier *Alluaudomyia melanosticta* (Ingram & Macfie) *Alluaudomyia mouensis* Giles & Wirth *Alluaudomyia neocaledoniensis* Clastrier *Alluaudomyia poguei* Giles & Wirth *Alluaudomyia tillierorum* Clastrier	Ceratopogonidae, Diptera
Alouattalges	Psoroptidae, Acari
Alouattamyia *Alouattamyia baeri* (Shannon & Greene)	Ceterebridae, Diptera
Alphitobius *Alphitobius diaperinus* (Panzer) *Alphitobius laevigatus* (Fabricius)	Tenebrionidae, Coleoptera
Alveonasus *Alveonasus lahorensis* (Neumann)	**see** *Ornithodoros* **see** *Ornithodoros lahorensis*
Alycus *Alycus roseus* Koch	Bimichaeliidae, Acari
Alysia *Alysia manducator* (Panzer)	Braconidae, Hymenoptera

Amalaraeus
Amalaraeus dissimilis (Jordan)
Amalaraeus penicilliger dissimilis (Jordan)
Amalaraeus penicilliger kratochvili (Rosický)
Amalaraeus penicilliger penicilliger (Grube)

Ceratophyllidae, Siphonaptera

see *Amalaraeus dissimilis*

Amaradix
Amaradix euphorbi (Rothschild)

Ceratophyllidae, Siphonaptera

Amblyomma
Amblyomma americanum (Linnaeus)
Amblyomma arianae Keirans & Garris
Amblyomma astrion Dönitz
Amblyomma babirussae Schulze
Amblyomma cajennense (Fabricius)
Amblyomma calcaratum Neumann
Amblyomma cohaerens Dönitz
Amblyomma cordiferum Neumann
Amblyomma cyprium Neumann
Amblyomma dissimile Koch
Amblyomma furcula Dönitz
Amblyomma gemma Dönitz
Amblyomma hebraeum Koch
Amblyomma helvolum Koch
Amblyomma incisum Neumann
Amblyomma inornatum (Banks)
Amblyomma integrum Karsch
Amblyomma lepidum Dönitz
Amblyomma limbatum Neumann
Amblyomma loculosum Neumann
Amblyomma maculatum Koch
Amblyomma marmoreum Koch
Amblyomma neumanni Ribaga
Amblyomma nodosum Neumann
Amblyomma nuttalli Dönitz
Amblyomma ovale Koch
Amblyomma parvum Aragão
Amblyomma paulopunctatum Neumann
Amblyomma pomposum Dönitz
Amblyomma rotundatum Koch
Amblyomma sparsum Neumann
Amblyomma testudinarium Koch
Amblyomma tholloni Neumann
Amblyomma tigrinum Koch
Amblyomma triguttatum Koch
Amblyomma variegatum (Fabricius)

Ixodidae, Acari

see *Amblyomma neumanni*

Amblyopinus
Amblyopinus tiptoni Barrera

Staphylinidae, Coleoptera

Ameletus
Ameletus alexandrae Brodsky

Siphlonuridae, Ephemeroptera

Ameroseius
Ameroseius corniculus Karg
Ameroseius plumigerus (Oudemans)

Ameroseiidae, Acari

Amiota
Amiota variegata (Fallén)

Drosophilidae, Diptera

Amorphacarus
Amorphacarus hengererorum Jameson
Amorphacarus yezoensis Lukoschus, Ono & Uchikawa

Myobiidae, Acari

Amphalius *Amphalius runatus* (Jordan & Rothschild)	Ceratophyllidae, Siphonaptera
Amphipsyche *Amphipsyche scottae* Kimmins	Hydropsychidae, Trichoptera
Amphipsylla *Amphipsylla longispina gongheensis* Zhang & Ma *Amphipsylla longispina longispina* Scalon	Leptopsyllidae, Siphonaptera
Ampulex *Ampulex compressa* (Fabricius)	Sphecidae, Hymenoptera
Amydria *Amydria selvae* Davis	Tineidae, Lepidoptera
Amyrsidea *Amyrsidea megalosoma* (Overgaard) *Amyrsidea perdicis megalosoma* (Overgaard) *Amyrsidea perdicis perdicis* (Denny) *Amyrsidea steineri* Price & Emerson	Somaphantidae, Phthiraptera **see** *Amyrsidea perdicis megalosoma*
Anabas *Anabas testudineus* (Bloch)	Anabantidae, Perciformes
Anabrus *Anabrus simplex* Haldeman	Tettigoniidae, Orthoptera
Analges *Analges chelopus* (Hermann)	Analgidae, Acari
Ananteris *Ananteris festae* Borelli *Ananteris luciae* Lourenço	Buthidae, Scorpiones
Anaphe *Anaphe panda* (Boisduval) *Anaphe venata* Butler	Thaumetopoeidae, Lepidoptera
Anaphlebotomus	**see** *Phlebotomus*
Anastatus *Anastatus bifasciatus* (Geoffroy) *Anastatus charitos* De Santis *Anastatus coreophagus* Ashmead *Anastatus disparis* Ruschka *Anastatus excavatus* (De Santis) *Anastatus gastropachae* Ashmead *Anastatus japonicus* Ashmead *Anastatus umae* Bouček	Eupelmidae, Hymenoptera **see** *Anastatus japonicus*
Anaticola *Anaticola anseris* (Linnaeus) *Anaticola crassicornis* Scopoli	Philopteridae, Phthiraptera
Anatoecus *Anatoecus dentatus* (Scopoli)	Philopteridae, Phthiraptera
Anatopynia *Anatopynia pennipes* Freeman	Chironomidae, Diptera
Ancala *Ancala necopina* (Austen)	Tabanidae, Diptera

Ancistrocerus *Ancistrocerus adiabatus* (Saussure) *Ancistrocerus antilope* (Panzer)	Eumenidae, Hymenoptera
Ancistroplax *Ancistroplax chodsigoae* Chin *Ancistroplax crocidurae* Waterston	Haematopinoididae, Phthiraptera
Andalgalomacarus *Andalgalomacarus paraguayensis* Goff & Whitaker	Trombiculidae, Acari
Androctonus *Androctonus aeneas* Koch *Androctonus amoreuxi* (Audouin & Savigny) *Androctonus australis australis* (Linnaeus) *Androctonus australis garzonii* Goyffon & Lamy *Androctonus australis hector* Koch *Androctonus crassicauda* (Olivier) *Androctonus mauretanicus* (Pocock)	Buthidae, Scorpiones
Androlaelaps *Androlaelaps benedictae* Fain & Hart *Androlaelaps casalis* (Berlese) *Androlaelaps fahrenholzi* (Berlese) *Androlaelaps geomys* Strandtmann *Androlaelaps glasgowi* Ewing *Androlaelaps guimaraesi* Fonseca *Androlaelaps kivuensis* Fain & Hart *Androlaelaps longipes* (Bregetova) *Androlaelaps pachyptilae* (Zumpt & Till) *Androlaelaps rahmi* Fain & Hart *Androlaelaps zuluensis* (Zumpt)	Laelapidae, Acari **see** *Androlaelaps fahrenholzi*
Anilocra *Anilocra capensis* Leach	Cymothoidae, Isopoda
Anisops *Anisops bouvieri* Kirkaldy *Anisops breddeni* Kirkaldy *Anisops deanei* Brooks *Anisops sardea* Herrich-Schäffer	Notonectidae, Hemiptera
Anisopus *Anisopus cinctus* (Fabricius) *Anisopus fenestralis* (Scopoli)	**see** *Sylvicola* **see** *Sylvicola cinctus* **see** *Sylvicola fenestralis*
Ankylohelea *Ankylohelea montana* Meillon & Wirth	Ceratopogonidae, Diptera
Anocentor *Anocentor nitens* (Neumann)	Ixodidae, Acari
Anoetus *Anoetus feroniarum* (Dufour)	Anoetidae, Acari
Anomala *Anomala exoleta* Faldermann *Anomala luculenta smaragdina* Ohaus	Scarabaeidae, Coleoptera
Anomalohimalaya *Anomalohimalaya lama* Hoogstraal, Kaiser & Mitchell *Anomalohimalaya lotozkyi* Filippova & Panova	Ixodidae, Acari

Anomiopsyllus
Anomiopsyllus amphibolus Wagner

Hystrichopsyllidae, Siphonaptera

Anopheles   Culicidae, Diptera
Anopheles aconitus Dönitz
Anopheles albimanus Wiedemann
Anopheles albitarsis Lynch Arribálzaga
Anopheles albitarsis domesticus Galvão & Damasceno   **see** *Anopheles marajoara*
Anopheles algeriensis Theobald
Anopheles allopha (Lutz & Peryassú)
Anopheles amazonicus Christophers   **see** *Anopheles mattogrossensis*
Anopheles amictus Edwards
Anopheles annularis van der Wulp
Anopheles annulipes Walker
Anopheles anthropophagus Xu & Feng   **see** *Anopheles lesteri anthropophagus*
Anopheles apicimacula Dyar & Knab
Anopheles aquasalis Curry
Anopheles arabicus Christophers & Khazan Chand   **see** *Anopheles fluviatilis*
Anopheles arabiensis Patton
Anopheles argyropus Swellengrebel
Anopheles atroparvus Van Thiel
Anopheles atropos Dyar & Knab
Anopheles balabacensis Baisas
Anopheles balabacensis dirus Peyton & Harrison   **see** *Anopheles dirus*
Anopheles bancroftii Giles
Anopheles barberi Coquillett
Anopheles barbirostris van der Wulp
Anopheles barbumbrosus Strickland & Chowdhury
Anopheles beklemishevi Stegniǐ & Kabanova
Anopheles bellator Dyar & Knab
Anopheles benarrochi Gabaldon *et al.*
Anopheles boliviensis (Theobald)
Anopheles bradleyi King
Anopheles braziliensis (Chagas)
Anopheles brunnipes (Theobald)
Anopheles bwambae White
Anopheles campestris Reid
Anopheles changfus Ma
Anopheles claviger (Meigen)
Anopheles collessi Reid
Anopheles coustani Laveran
Anopheles crawfordi Reid
Anopheles crucians Wiedemann
Anopheles cruzii Dyar & Knab
Anopheles culicifacies Giles
Anopheles cydippis De Meillon
Anopheles darlingi Root
Anopheles demeilloni Evans
Anopheles dirus Peyton & Harrison
Anopheles domesticus Galvão & Damasceno   **see** *Anopheles marajoara*
Anopheles donaldi Reid
Anopheles dravidicus Christophers
Anopheles dthali Patton
Anopheles earlei Vargas
Anopheles elegans (James)
Anopheles engarensis Kanda & Oguma
Anopheles ethiopicus Gillies & Coetzee
Anopheles evansi (Brèthes)
Anopheles farauti Laveran
Anopheles flavirostris (Ludlow)
Anopheles fluviatilis James
Anopheles franciscanus McCracken   **see** *Anopheles pseudopunctipennis*
franciscanus
Anopheles freeborni Aitken

Anopheles funestus Giles
Anopheles galvaoi Causey, Deane & Deane
Anopheles gambiae Giles
Anopheles hancocki Edwards
Anopheles hargreavesi Evans
Anopheles hermsi Barr & Guptavanij
Anopheles hilli Woodhill & Lee
Anopheles hispaniola (Theobald)
Anopheles hyrcanus (Pallas)
Anopheles hyrcanus sinensis Wiedemann   **see** *Anopheles sinensis*
Anopheles indefinitus (Ludlow)
Anopheles insulaeflorum (Swellengrebel *et al.*)
Anopheles intermedius (Chagas)
Anopheles introlatus Colless
Anopheles jamesii Theobald
Anopheles jeyporiensis James
Anopheles jeyporiensis candidiensis Koidzumi   **see** *Anopheles jeyporiensis*
Anopheles kochi Dönitz
Anopheles koliensis Owen
Anopheles koreicus Yamada & Watanabe
Anopheles kosiensis Coetzee, Segerman & Hunt
Anopheles kunmingensis Dong & Wang
Anopheles kweiyangensis Yao & Wu
Anopheles labranchiae labranchiae Falleroni
Anopheles labranchiae race *sicaulti* Roubaud
Anopheles lanei Galvão & Amaral
Anopheles lepidotus Zavortink
Anopheles lesteri anthropophagus Xu & Feng
Anopheles lesteri lesteri Baisas & Hu
Anopheles letifer Sandosham
Anopheles leucosphyrus Dönitz
Anopheles lindesayi japonicus Yamada
Anopheles litoralis King
Anopheles lounibosi Gillies & Coetzee
Anopheles ludlowae (Theobald)
Anopheles maculatus Theobald
Anopheles maculipennis Meigen
Anopheles maculipennis atroparvus Van Thiel   **see** *Anopheles atroparvus*
Anopheles maculipennis messeae Falleroni   **see** *Anopheles messeae*
Anopheles maculipennis sacharovi Favr   **see** *Anopheles sacharovi*
Anopheles maculipennis subalpinus Hackett & Lewis   **see** *Anopheles subalpinus*
Anopheles maculipes (Theobald)
Anopheles mangyanus (Banks)
Anopheles marajoara Galvão & Damasceno
Anopheles marshallii (Theobald)
Anopheles martinius Shingarev
Anopheles mascarensis De Meillon
Anopheles mattogrossensis Lutz & Neiva
Anopheles mediopunctatus (Theobald)
Anopheles melanoon Hackett
Anopheles melanoon subalpinus Hackett & Lewis   **see** *Anopheles subalpinus*
Anopheles melas Theobald
Anopheles merus Dönitz
Anopheles messeae Falleroni
Anopheles minimus Theobald
Anopheles moucheti Evans
Anopheles mousinhoi De Meillon & de Carvalho Pereira
Anopheles multicolor Cambouliu
Anopheles namibiensis Coetzee
Anopheles neivai Howard, Dyar & Knab
Anopheles nemophilous Peyton & Ramalingam
Anopheles nigerrimus Giles
Anopheles nili (Theobald)
Anopheles nitidus Harrison, Scanlon & Reid

Anopheles *nivipes* (Theobald)
Anopheles *noroestensis* Galvão & Lane   **see** *Anopheles evansi*
Anopheles *notanandai* Rattanarithikul & Green
Anopheles *nuneztovari* Gabaldon
Anopheles *occidentalis* Dyar & Knab
Anopheles *oswaldoi* (Peryassú)
Anopheles *pallidus* Theobald
Anopheles *paltrinierii* Shidrawi & Gillies
Anopheles *paludis* Theobald
Anopheles *pattoni* Christophers
Anopheles *pauliani* Grjebine
Anopheles *peditaeniatus* (Leicester)
Anopheles *perplexens* Ludlow
Anopheles *peryassui* Dyar & Knab
Anopheles *petragnanii* Del Vecchio
Anopheles *peytoni* Kulasekera, Harrison & Amerasinghe
Anopheles *pharoensis* Theobald
Anopheles *philippinensis* Ludlow
Anopheles *pictus* Loew   **see** *Anopheles hyrcanus*
Anopheles *pitchfordi* (Giles)   **see** *Anopheles marshallii*
Anopheles *plumbeus* Stephens
Anopheles *pretoriensis* (Theobald)
Anopheles *pseudojamesi* Strickland & Chowdhury
Anopheles *pseudopunctipennis boydi* Vargas
Anopheles *pseudopunctipennis franciscanus* McCracken
Anopheles *pseudopunctipennis pseudopunctipennis* Theobald
Anopheles *pseudowillmori* (Theobald)
Anopheles *pulcherrimus* Theobald
Anopheles *punctimacula* Dyar & Knab
Anopheles *punctipennis* (Say)
Anopheles *punctulatus* Dönitz
Anopheles *quadriannulatus davidsoni* Ribeiro *et al.*
Anopheles *quadriannulatus quadriannulatus* (Theobald)
Anopheles *quadrimaculatus* Say
Anopheles *ramsayi* Covell   **see** *Anopheles pseudojamesi*
Anopheles *rangeli* Gabaldon *et al.*
Anopheles *rivulorum* Leeson
Anopheles *rufipes* (Gough)
Anopheles *sacharovi* Favr
Anopheles *salbaii* Maffi & Coluzzi
Anopheles *sawadwongporni* Rattanarithikul & Green
Anopheles *sawyeri* Causey, Deane, Deane & Sampaio
Anopheles *sergentii* (Theobald)
Anopheles *sicaulti* Roubaud   **see** *Anopheles labranchiae* race *sicaulti*
Anopheles *sinensis* Wiedemann
Anopheles *sineroides* Yamada
Anopheles *splendidus* Koidzumi
Anopheles *squamifemur* Antunes
Anopheles *squamosus* Theobald
Anopheles *stephensi* Liston
Anopheles *stephensi* var. *mysorensis* Sweet & Rao   **see** *Anopheles stephensi*
Anopheles *strodei* Root
Anopheles *subalpinus* Hackett & Lewis
Anopheles *subpictus* Grassi
Anopheles *sulawesi* Waktoedi
Anopheles *sundaicus* (Rodenwaldt)
Anopheles *superpictus* Grassi
Anopheles *takasagoensis* Morishita
Anopheles *tenebrosus* Dönitz
Anopheles *tessellatus* Theobald
Anopheles *theobaldi* Giles
Anopheles *triannulatus* (Neiva & Pinto)
Anopheles *trinkae* Faran
Anopheles *turkhudi* Liston

Anopheles typicus Hackett & Missiroli — **see** Anopheles maculipennis maculipennis
Anopheles umbrosus (Theobald)
Anopheles vagus Dönitz
Anopheles vaneedeni Gillies & Coetzee
Anopheles varuna Iyengar
Anopheles venezuelae Evans — **see** Anopheles punctimacula
Anopheles vestitipennis Dyar & Knab
Anopheles walkeri Theobald
Anopheles wellcomei Theobald
Anopheles willmori (James)
Anopheles yatsushiroensis Miyazaki
Anopheles ziemanni Grünberg

Anoplotrupes — **see** Geotrupes
Anoplotrupes stercorosus (Scriba) — **see** Geotrupes stercorosus

Antarctophthirus — Echinophthiriidae, Phthiraptera
Antarctophthirus ogmorhini Enderlein

Antennoseius — Ascidae, Acari
Antennoseius bregetovae Chelebiev
Antennoseius koroljevae Chelebiev

Anthia — Carabidae, Coleoptera
Anthia sexguttata (Fabricius)

Anthocoris — Anthocoridae, Hemiptera
Anthocoris nemorum (Linnaeus)

Anthomyia — Anthomyiidae, Diptera
Anthomyia illocata Walker

Anthrenus — Dermestidae, Coleoptera
Anthrenus flavipes LeConte

Anystis — Anystidae, Acari
Anystis baccarum (Linnaeus)
Anystis jabanica (Berlese)
Anystis salicinus (Linnaeus)

Aparapotamon — Sinopotamidae, Decapoda

Apatolestes — Tabanidae, Diptera
Apatolestes actites Philip & Steffan

Aphaereta — Braconidae, Hymenoptera
Aphaereta aotea Hughes & Woolcock
Aphaereta pallipes (Say)

Aphelomyia — **see** Sarcodexiopsis
Aphelomyia welchi (Hall) — **see** Sarcodexiopsis welchi

Aphodius — Scarabaeidae, Coleoptera
Aphodius aenictus Cooper & Gordon
Aphodius agullasi Endrődi
Aphodius aprilis Bordat
Aphodius ater (DeGeer)
Aphodius atsushii Ochi
Aphodius caprai Dellacasa
Aphodius citellorum Semenov-Tian-Shanskiĭ & Medvedev
Aphodius contaminatus (Herbst)
Aphodius coreensis Kim
Aphodius culminarius Reitter

Aphodius dellacasai Avila
Aphodius distinctus (O. F. Müller)
Aphodius erraticus (Linnaeus)
Aphodius erytoides Endrődi
Aphodius fimetarius (Linnaeus)
Aphodius fossor (Linnaeus)   **see** *Colobopterus fossor*
Aphodius ganabi Endrődi
Aphodius granarius (Linnaeus)
Aphodius haemorrhoidalis (Linnaeus)
Aphodius hammondi Dellacasa
Aphodius hoberlandti Tesař
Aphodius holgati Endrődi
Aphodius ibericus Harold
Aphodius impunctatus Waterhouse
Aphodius karrooensis Endrődi
Aphodius koreanensis Kim
Aphodius lividus (Olivier)
Aphodius luridus (Fabricius)
Aphodius pittinoi Carpaneto
Aphodius pongensis Endrődi
Aphodius psammophilus Endrődi   **see** *Aphodius ganabi*
Aphodius rectus (Motschulsky)
Aphodius reyi Reitter
Aphodius rufipes (Linnaeus)
Aphodius rufus (Moll)
Aphodius scybalarius (Fabricius)   **see** *Aphodius rufus*
Aphodius silvestris Bordat
Aphodius stefenellii Dellacasa
Aphodius suarius Faldermann
Aphodius sublimbatus (Motschulsky)
Aphodius subterraneus krasnojarskicus Dellacasa
Aphodius subterraneus subterraneus (Linnaeus)
Aphodius tasmaniae Hope
Aphodius transaralicus Nikolaev
Aphodius urostigma Harold

Aphodoharmogaster   Scarabaeidae, Coleoptera
Aphodoharmogaster costata Endrődi

Aphonopelma   Theraphosidae, Araneae
Aphonopelma chalcodes Chamberlin   **see** *Rhechostica chalcodes*
Aphonopelma seemanni (F. Pickard–Cambridge)   **see** *Rhechostica seemanni*

Aphrania   Cimicidae, Hemiptera
Aphrania recta Ferris & Usinger

Aphrophora   Aphrophoridae, Hemiptera
Aphrophora alni (Fallén)

Apis   Apidae, Hymenoptera
Apis cerana cerana Fabricius
Apis cerana indica Fabricius
Apis dorsata Fabricius
Apis florea Fabricius
Apis indica Fabricius   **see** *Apis cerana indica*
Apis mellifera ligustica Spinola
Apis mellifera mellifera Linnaeus
Apis mellifera scutellata Lepeletier
Apis mellifica Linnaeus   **see** *Apis mellifera*

Apodochondria   Chondracanthidae, Copepoda
Apodochondria medusae Ho & Dojiri

Aponomma
Aponomma exornatum (Koch)
Aponomma flavomaculatum (Lucas)
Aponomma hydrosauri (Denny)
Aponomma latum (Koch)
Aponomma lucasi Warburton
Aponomma transversale (Lucas)
Aponomma varanensis (Supino)

Ixodidae, Acari

Aporoptychus
Aporoptychus lamottei Dresco

Ctenizidae, Araneae

Aprostocetus
Aprostocetus asthenogmus (Waterston)
Aprostocetus hagenowii (Ratzeburg)

Eulophidae, Hymenoptera

Araeomerus

Hemimeridae, Dermaptera

Aralichus
Aralichus ambiguae Atyeo
Aralichus anodorhynchi Atyeo
Aralichus araraunae Atyeo
Aralichus canestrinii (Trouessart)
Aralichus chloropterae Atyeo
Aralichus couloni Atyeo
Aralichus elongatus Pérez & Atyeo
Aralichus manilatae Atyeo
Aralichus maracanae Atyeo
Aralichus menchacai Pérez & Atyeo
Aralichus mexicanus Atyeo
Aralichus militaris Atyeo
Aralichus nobilis Atyeo
Aralichus severae Atyeo

Pterolichidae, Acari

Araneus
Araneus diadematus Clerck
Araneus gemma (McCook)

Araneidae, Araneae

Archaeopsylla
Archaeopsylla erinacei (Bouché)

Pulicidae, Siphonaptera

Archimandrita

Blattidae, Dictyoptera

Ardeicola
Ardeicola expallidus Blagoveshchenskiĭ

Philopteridae, Phthiraptera

Arenivaga

Polyphagidae, Dictyoptera

Argas
Argas africolumbae Hoogstraal et al.
Argas arboreus Kaiser, Hoogstraal & Kohls
Argas beijingensis Teng
Argas brumpti Neumann
Argas columbarum (Shaw)
Argas confusus Hoogstraal
Argas cooleyi Kohls & Hoogstraal
Argas gilcolladoi Estrada-Peña et al.
Argas japonicus Yamaguti, Clifford & Tipton
Argas macrostigmatus Filippova
Argas miniatus Koch
Argas persicus (Oken)
Argas polonicus Siuda, Hoogstraal, Clifford & Wassef
Argas pusillus Kohls
Argas radiatus Railliet

Argasidae, Acari

Argas reflexus (Fabricius)
Argas robertsi Hoogstraal, Kaiser & Kohls
Argas sinensis Jeu & Zhu
Argas streptopelia Kaiser, Hoogstraal & Horner
Argas theilerae Hoogstraal & Kaiser
Argas vespertilionis (Latreille)
Argas vulgaris Filippova
Argas walkerae Kaiser & Hoogstraal
Argas zumpti Hoogstraal, Kaiser & Kohls

Arge   Argidae, Hymenoptera
Arge pullata (Zaddach)

Argia   Coenagriidae, Odonata
Argia fumipennis (Burmeister)

Argiope   Araneidae, Araneae
Argiope aurantia Lucas
Argiope lobata (Pallas)

Argitus   Macronyssidae, Acari
Argitus oryzomys Yunker & Saunders

Argulus   Argulidae, Branchiura
Argulus africanus Thiele
Argulus alosae Gould
Argulus canadensis Wilson
Argulus chesapeakensis Cressey
Argulus siamensis Wilson

Argyope   **see** *Argiope*

Argyroneta   Argyronetidae, Araneae
Argyroneta aquatica (Clerck)

Arilus   Reduviidae, Hemiptera
Arilus cristatus (Linnaeus)

Armigeres   Culicidae, Diptera
Armigeres annulitarsis (Leicester)
Armigeres dolichocephalus (Leicester)
Armigeres durhami (Edwards)
Armigeres milnensis Lee
Armigeres obturbans (Walker)
Armigeres pectinatus (Edwards)
Armigeres subalbatus (Coquillett)

Armillifer   Armilliferidae, Pentastomida
Armillifer armillatus (Wyman)
Armillifer moniliformis (Diesing)

Arrenurus   Arrenuridae, Acari
Arrenurus angustilimbatus Mullen
Arrenurus danbyensis Mullen
Arrenurus kenki Marshall
Arrenurus novimarshallae Wilson
Arrenurus pseudotenuicollis Wilson

Arthroceras   Xylophagidae, Diptera
Arthroceras fulvicorne Nagatomi
Arthroceras fulvicorne nigricapite Nagatomi   **see** *Arthroceras fulvicorne*
Arthroceras fulvicorne subsolanum Nagatomi   **see** *Arthroceras fulvicorne*
Arthroceras leptis (Osten Sacken)
Arthroceras pollinosum Williston

Ascoschoengastia
Ascoschoengastia balcanica (Kolebinova)
Ascoschoengastia brachytrichia (Brennan)
Ascoschoengastia brennani (Goff, Whitaker & Dietz)
Ascoschoengastia hoplodactyla (Goff, Loomis & Ainsworth)
Ascoschoengastia indica (Hirst)
Ascoschoengastia menghaiensis Yu, Yang & Gong
Ascoschoengastia nanjiangensis Zhou, Chen & Wang
Ascoschoengastia nicaraguae (Webb & Loomis)

Trombiculidae, Acari

Aspidoptera

Streblidae, Diptera

Astylus
Astylus atromaculatus (Blanchard)

Melyridae, Coleoptera

Ataenius

Scarabaeidae, Coleoptera

Atergatis
Atergatis floridus (Linnaeus)

Xanthidae, Decapoda

Atherigona
Atherigona orientalis Schiner

Muscidae, Diptera

Atholus
Atholus bimaculatus (Linnaeus)
Atholus rothkirchi Bickhardt

Histeridae, Coleoptera

Athripsodes

Leptoceridae, Trichoptera

Atractomorpha
Atractomorpha crenulata (Fabricius)

Acrididae, Orthoptera

Atrax
Atrax formidabilis Rainbow
Atrax infensus Hickman
Atrax robustus Pickard-Cambridge
Atrax versutus Rainbow

Dipluridae, Araneae

Atricholaelaps
Atricholaelaps glasgowi (Ewing)
Atricholaelaps guimaraesi (Fonseca)

see *Androlaelaps*
see *Androlaelaps fahrenholzi*
see *Androlaelaps guimaraesi*

Atrichopogon
Atrichopogon wirthi Chan & Linley

Ceratopogonidae, Diptera

Atta
Atta bisphaerica Forel
Atta capiguara Gonçalves
Atta laevigata (F. Smith)
Atta vollenweideri Forel

Formicidae, Hymenoptera

Attagenus
Attagenus unicolor (Brahm)

Dermestidae, Coleoptera

Atylotus
Atylotus agrestis (Wiedemann)
Atylotus bivittateinus Takahasi
Atylotus flavoguttatus (Szilady)
Atylotus kakeromaensis Hayakawa, Takahasi & Suzuki
Atylotus miser (Szilady)
Atylotus ozensis Hayakawa
Atylotus pulchellus karybenthinus (Szilady)
Atylotus pulchellus pulchellus (Loew)
Atylotus takaraensis Hayakawa & Takahasi

Tabanidae, Diptera

Atylotus thoracicus (Hine)

Atyphloceras
Atyphloceras multidentatus (Fox)
Atyphloceras nuperus (Jordan)

Hystrichopsyllidae, Siphonaptera

Aulacophyto
Aulacophyto baumgartneri Tibana & Souza Lopes
Aulacophyto reinhardi Tibana & Souza Lopes
Aulacophyto rusca Hall
Aulacophyto tarmaensis Tibana & Souza Lopes

Sarcophagidae, Diptera

Austracarus
Austracarus lukoschusi Goff
Austracarus masonae Goff

Leeuwenhoekiidae, Acari

Austroconops
Austroconops mcmillani Wirth & Lee

Ceratopogonidae, Diptera

Austrosimulium
Austrosimulium australense (Schiner)
Austrosimulium bancrofti (Taylor)
Austrosimulium colboi Davies & Györkös
Austrosimulium laticorne Tonnoir
Austrosimulium multicorne Tonnoir

Simuliidae, Diptera

Auyantepuia

see *Broteochactas*

Avicularia
Avicularia californicum (Ausserer)

Theraphosidae, Araneae

Azelia
Azelia cilipes (Haliday)

Muscidae, Diptera

B
Baculum
Baculum extradentatum (Brunner)

Phasmatidae, Phasmida

Badumna
Badumna insignis (L. Koch)

Desidae, Araneae

Bakerdania
Bakerdania plurisetosa Mahunka

Pygmephoridae, Acari

Bakericheyla
Bakericheyla chanayi (Berlese & Trouessart)

Cheyletidae, Acari

Balaustium
Balaustium murorum (Hermann)

Erythraeidae, Acari

Bardistus
Bardistus cibarius Newman

Cerambycidae, Coleoptera

Bareogonalos
Bareogonalos jezoensis (Uchida)

Trigonalidae, Hymenoptera

Barreropsylla
Barreropsylla excelsa Jordan

Stephanocircidae, Siphonaptera

Barypeithes
Barypeithes pellucidus (Boheman)

Curculionidae, Coleoptera

Basilia
Basilia bathybothyra Speis

Nycteribiidae, Diptera

Basilia limbella Maa
Basilia mediterranea Hurka
Basilia mongolensis mongolensis Theodor
Basilia mongolensis nudior Hurka
Basilia roylii (Westwood)

Bathyergolichus   Atopomelidae, Acari
Bathyergolichus hottentotus Fain

Bdellonyssus   **see** *Ornithonyssus*
Bdellonyssus bacoti (Hirst)   **see** *Ornithonyssus bacoti*
Bdellonyssus bursa (Berlese)   **see** *Ornithonyssus bursa*

Bdellorhynchus   Dermoglyphidae, Acari
Bdellorhynchus psalidurus Trouessart   **see** *Zygochelifer psalidurus*

Bellieria   Sarcophagidae, Diptera
Bellieria melanura (Meigen)   **see** *Helicophagella melanura*

Bellieriomima   Sarcophagidae, Diptera
Bellieriomima yaanensis Feng

Belminus   Reduviidae, Hemiptera

Belonogaster   Vespidae, Hymenoptera
Belonogaster petiolata (DeGeer)

Belostoma   Belostomatidae, Hemiptera
Belostoma boscii Lepeletier & Serville
Belostoma flumineum Say
Belostoma indicum Lepeletier & Serville   **see** *Lethocerus indicus*

Bembix   Sphecidae, Hymenoptera
Bembix americana americana Fabricius
Bembix americana comata Parker
Bembix antoni Krombein & Vecht
Bembix borrei Handlirsch
Bembix glauca Fabricius
Bembix orientalis Handlirsch

Bercaea   **see** *Sarcophaga*
Bercaea cruentata (Meigen)   **see** *Sarcophaga cruentata*
Bercaea haemorrhoidalis (Fallén)   **see** *Sarcophaga cruentata*

Besnoitia   Sarcocystidae, Eucoccidiorida
Besnoitia besnoiti (Marotel)

Bewsiella   Macronyssidae, Acari

Bezzia   Ceratopogonidae, Diptera
Bezzia calceata (Walker)
Bezzia lophophora Clastrier
Bezzia picticornis (Kieffer)
Bezzia xanthocephala Goetghebuer

Blaberolaelaps   Laelapidae, Acari
Blaberolaelaps beckeri Hunter, Rosario & Flechtmann

Blaberus   Blaberidae, Dictyoptera
Blaberus craniifer Burmeister
Blaberus discoidalis Serville
Blaberus fuscus Brunner   **see** *Blaberus craniifer*
Blaberus giganteus (Linnaeus)
Blaberus trapezoideus Burmeister

Blaesoxipha *Blaesoxipha aspinata* (Senior White)	Sarcophagidae, Diptera
Blankaartia *Blankaartia acuscutellaris* (Walch)	Trombiculidae, Acari
Blaps *Blaps lethifera* Marsh	Tenebrionidae, Coleoptera
Blaptica *Blaptica dubia* Serville	Blaberidae, Dictyoptera
Blatta *Blatta furcata* (Karny) *Blatta lateralis* (Walker) *Blatta orientalis* Linnaeus *Blatta transfuga* Brünnich	Blattidae, Dictyoptera **see** *Blattella germanica*
Blattella *Blattella asahinai* Mizukubo *Blattella beybienkoi* Roth *Blattella germanica* (Linnaeus) *Blattella lecordieri* (Princis) *Blattella lituricollis* (Walker) *Blattella orientalis* (Linnaeus) *Blattella sordida* (Shelford) *Blattella vaga* Hebard	Blattellidae, Dictyoptera **see** *Blattella asahinai* **see** *Blatta orientalis*
Blattisocius *Blattisocius keegani* Fox *Blattisocius mali* (Oudemans)	Ascidae, Acari
Bledius *Bledius mandibularis* Gyllenhal *Bledius spectabilis* Kraatz	Staphylinidae, Coleoptera
Blomia *Blomia freemani* Hughes *Blomia thori* Zakhvatkin *Blomia tropicalis* van Bronswijk, de Cock & Oshima	Glycyphagidae, Acari
Boettcherisca *Boettcherisca formosensis* Kirner & Lopes *Boettcherisca invaria* (Walker) *Boettcherisca javanica* Lopes *Boettcherisca karnyi* (Hardy) *Boettcherisca koimani* Kano & Shinonaga *Boettcherisca nathani* Lopes *Boettcherisca peregrina* (Robineau-Desvoidy) *Boettcherisca septentrionalis* Rodendorf *Boettcherisca timorensis* Kano & Shinonaga	**see** *Sarcophaga* **see** *Sarcophaga formosensis* **see** *Sarcophaga invaria* **see** *Sarcophaga javanica* **see** *Sarcophaga karnyi* **see** *Sarcophaga koimani* **see** *Sarcophaga nathani* **see** *Sarcophaga peregrina* **see** *Sarcophaga septentrionalis* **see** *Sarcophaga timorensis*
Bolbelasmus	Geotrupidae, Coleoptera
Bolbocerosoma *Bolbocerosoma nigroplagiatum* (Waterhouse)	Scarabaeidae, Coleoptera
Bolbodera	Reduviidae, Hemiptera
Bombus *Bombus agrorum* Fabricius *Bombus pascuorum* Scopoli *Bombus pennsylvanicus* (DeGeer)	Apidae, Hymenoptera **see** *Bombus pascuorum*

Bomolochus *Bomolochus cuneatus* Fraser	Bomolochidae, Copepoda
Bonomiella *Bonomiella columbae* Emerson	Menoponidae, Phthiraptera
Bonomoia	Anoetidae, Acari
Boophilus *Boophilus annulatus* (Say) *Boophilus annulatus calcaratus* (Birulya) *Boophilus calcaratus* (Birulya) *Boophilus decoloratus* (Koch) *Boophilus geigyi* Aeschlimann & Morel *Boophilus kohlsi* Hoogstraal & Kaiser *Boophilus microplus* (Canestrini) *Boophilus sharifi* Minning *Boophilus sharifi* Siddiqi & Jan	Ixodidae, Acari **see** *Boophilus annulatus* **see** *Boophilus annulatus* *nomen nudum*
Boophthora *Boophthora erythrocephala* (DeGeer)	**see** *Simulium* **see** *Simulium erythrocephalum*
Booponus *Booponus intonsus* Aldrich	Calliphoridae, Diptera
Borborillus *Borborillus frigipennis* (Spuler) *Borborillus singularis* (Spuler)	Sphaeroceridae, Diptera **see** *Copromyza frigipennis* **see** *Copromyza singularis*
Borborus *Borborus flavipennis* Haliday *Borborus hamatus* Haliday *Borborus longipennis* Haliday *Borborus suillorum* Haliday	**see** *Copromyza* **see** *Copromyza pallifrons* *nonem dubium* **see** *Copromyza vitripennis* **see** *Copromyza fimetaria*
Boreellus *Boreellus atriceps* (Zetterstedt)	Calliphoridae, Diptera
Boreocanthon *Boreocanthon ebenus* (Say)	**see** *Canthon* **see** *Canthon ebenus*
Bothamia *Bothamia demeilloni* Meiswinkel	Ceratopogonidae, Diptera
Botyodes *Botyodes principalis* Leech	Pyralidae, Lepidoptera
Bovicola *Bovicola americana* Jellison *Bovicola bovis* (Linnaeus) *Bovicola caprae* (Gurlt) *Bovicola crassipes* (Rudow) *Bovicola equi* (Denny) *Bovicola limbata* (Gervais) *Bovicola longicornis* (Nitzsch) *Bovicola ovis* (Schrank)	Trichodectidae, Phthiraptera **see** *Bovicola longicornis* **see** *Werneckiella equi*
Brachydeutera *Brachydeutera longipes* Hendel *Brachydeutera munroi* Cresson	Ephydridae, Diptera
Brachydiplax *Brachydiplax farinosa* Krüger	Libellulidae, Odonata

Brachymeria	Chalcididae, Hymenoptera
Brachymeria calliphorae (Froggatt)	**see** *Brachymeria ucalegon*
Brachymeria discreta Gahan	
Brachymeria discretoidea Gahan	**see** *Brachymeria discreta*
Brachymeria fonscolombei (Dufour)	**see** *Brachymeria podagrica*
Brachymeria lasus (Walker)	
Brachymeria minuta (Linnaeus)	
Brachymeria neglecta (Masi)	**see** *Brachymeria podagrica*
Brachymeria obscurata Walker	**see** *Brachymeria lasus*
Brachymeria podagrica (Fabricius)	
Brachymeria ucalegon (Walker)	
Brachypelma	Theraphosidae, Araneae
Brachypelma albopilosa Valerio	
Brachypelma emilia Smith	
Brachypelma smithi (F. Pickard-Cambridge)	
Brachypogon	Ceratopogonidae, Diptera
Brachypogon kokocinskii Szadziewski	
Brachypogon kremeri Szadziewski & Havelka	
Brachypogon krzeminskii Szadziewski & Havelka	
Brachypogon nieves (Havelka)	
Brachypogon pakistanicus Szadziewski & Havelka	
Brachypogon vitiosus (Winnertz)	
Brachysomus	Curculionidae, Coleoptera
Brachysomus echinatus (Bonsdorff)	
Brachytarsina	Streblidae, Diptera
Brachytarsina sinhai Vazirani & Advani	
Brachytarsina trinotata Maa	
Bracon	Braconidae, Hymenoptera
Bracon hebetor Say	
Bradysia	Sciaridae, Diptera
Bradysia coprophila (Lintner)	
Brennanacarus	Trombiculidae, Acari
Brennanacarus annereauxi (Brennan & Yunker)	
Brennania	Tabanidae, Diptera
Brennania hera Osten Sacken	
Brontaea	Muscidae, Diptera
Broteochactas	Chactidae, Scorpiones
Broteochactas gollmeri (Karsch)	
Broteochactas laui Kjellesvig-Waering	
Broteochactas nitidus Pocock	
Brueelia	Philopteridae, Phthiraptera
Brueelia abluda (Złotorzycka)	
Brueelia amsel (Eichler)	
Brueelia merulensis (Denny)	
Brumptomyia	Psychodidae, Diptera
Brumptomyia bragai Mangabeira & Sherlock	
Brumptomyia guimaraesi (Coutinho & Barretto)	
Brumptomyia pintoi (Costa Lima)	
Brumptomyia troglodytes (Lutz)	
Bryobia	Tetranychidae, Acari
Bryobia cristata (Dugès)	

Bryobia praetiosa Koch

Bucimex Cimicidae, Hemiptera
Bucimex chilensis Usinger

Buthacus Buthidae, Scorpiones

Buthotus Buthidae, Scorpiones
Buthotus judaicus (Simon)
Buthotus minax (L. Koch)
Buthotus tamulus (Fabricius)

Buthus Buthidae, Scorpiones
Buthus crassicauda (Olivier) **see** *Androctonus crassicauda*
Buthus eupeus (Koch) **see** *Mesobuthus eupeus*
Buthus hendersoni Pocock
Buthus martensi Karsch
Buthus minax L. Koch **see** *Buthotus minax*
Buthus occitanus mardochei Simon
Buthus occitanus occitanus (Amoureux)
Buthus occitanus tunetanus (Herbst)
Buthus quinquestriatus (Hemprich & Ehrenberg) **see** *Leiurus quinquestriatus*
Buthus tamulus (Fabricius) **see** *Buthotus tamulus*

Byersalges Falculiferidae, Acari
Byersalges phyllophorus Gaud & Barré
Byersalges talpacoti (Černý)

Byrsotria Blattidae, Dictyoptera
Byrsotria fumigata (Guérin-Méneville)

C
Caccobius Scarabaeidae, Coleoptera
Caccobius jessoensis Harold
Caccobius ultor (Sharp)
Caccobius vulcanus (Fabricius)

Cacodmus Cimicidae, Hemiptera
Cacodmus indicus Jordan & Rothschild
Cacodmus sumatrensis Ferris & Usinger
Cacodmus villosus (Stål)

Cadrema Chloropidae, Diptera
Cadrema pallida pallida (Loew)
Cadrema pallida var. *bilineata* (de Meijere)

Caenopsylla Leptopsyllidae, Siphonaptera
Caenopsylla janineae Beaucournu & Gouat

Calamicoptes Laminosioptidae, Acari
Calamicoptes galli Lombert, Gaud & Lukoschus

Caligus Caligidae, Copepoda
Caligus elongatus Nordmann
Caligus infestans Heller
Caligus lacustris Steenstrup & Lütken
Caligus pelamydis Krøyer
Caligus tenuifurcatus Wilson

Calilena Agelenidae, Araneae

Callidosoma Erythraeidae, Acari
Callidosoma guatemalensis Treat

Callinectes	Portunidae, Decapoda
Calliphora	Calliphoridae, Diptera
Calliphora augur (Fabricius)	
Calliphora croceipalpis Jaennicke	
Calliphora erythrocephala (Meigen)	**see** *Calliphora vicina*
Calliphora grahami Aldrich	**see** *Aldrichina grahami*
Calliphora hilli Patton	
Calliphora hortona (Walker)	**see** *Xenocalliphora hortona*
Calliphora lata Coquillett	**see** *Calliphora nigribarbis*
Calliphora nigribarbis Vollenhoven	
Calliphora nociva Hardy	**see** *Calliphora placida*
Calliphora placida Walker	
Calliphora quadrimaculata (Swederus)	
Calliphora stygia (Fabricius)	
Calliphora tibialis Macquart	**see** *Onesia tibialis*
Calliphora uralensis Villeneuve	
Calliphora vicina Robineau-Desvoidy	
Calliphora vomitoria (Linnaeus)	
Calliptamus	Acrididae, Orthoptera
Callitroga	**see** *Cochliomyia*
Callopsylla	Ceratophyllidae, Siphonaptera
Callopsylla beishanensis Wu, Ni & Wu	
Callopsylla caspia (Ioff & Argyropulo)	
Callopsylla digitata Cai, Wu & Liu	
Callopsylla gypaetina Peus	
Caloglyphus	Acaridae, Acari
Caloglyphus berlesei (Michael)	**see** *Caloglyphus neotomae*
Caloglyphus caroli ChannaBasavanna & Rao	
Caloglyphus neotomae (Fain & Whitaker)	
Caloglyphus paradoxus (Oudemans)	
Calyptra	Noctuidae, Lepidoptera
Calyptra eustrigata (Hampson)	
Calyptra fasciata (Moore)	
Calyptra labilis (Berio)	**see** *Calyptra fasciata*
Calyptra minuticornis minuticornis (Guenée)	
Calyptra minuticornis novaepommeraniae (Strand)	
Calyptra orthograpta (Butler)	
Campanulotes	Goniodidae, Phthiraptera
Campanulotes bidentatus bidentatus (Scopoli)	
Campanulotes bidentatus compar (Burmeister)	
Camponotus	Formicidae, Hymenoptera
Camponotus aegyptiacus Emery	
Camponotus brasiliensis Mayr	
Camponotus vagus (Scopoli)	
Campsomeris	Scoliidae, Hymenoptera
Campsomeris sexmaculata Fabricius	**see** *Colpa interrupta*
Camptochironomus	Chironomidae, Diptera
Camptochironomus tentans (Fabricius)	
Camptotypus	Ichneumonidae, Hymenoptera
Camptotypus pulchripennis (Saussure)	
Campyloneura	Miridae, Hemiptera
Campyloneura virgula (Herrich-Schäffer)	

Canthon	Scarabaeidae, Coleoptera
Canthon aberrans (Harold)	
Canthon ebenus Say	
Canthon indigaceus chevrolati Harold	
Canthon indigaceus indigaceus LeConte	
Canthon rugosum Blanchard	
Canthon viridis championi Bates	
Canthon viridis leechi (Martínez, Halffter & Halffter)	
Canthon viridis viridis (Palisot de Beauvois)	

Capehelea   Ceratopogonidae, Diptera
Capehelea steli Meillon & Wirth

Carastrum   Erythraeidae, Acari
Carastrum ferrari Southcott

Carausius   Phasmatidae, Phasmida
Carausius morosus Brunner von Wattenwyl

Carcinops   Histeridae, Coleoptera
Carcinops pumilio (Erichson)
Carcinops troglodytes (Paykull)

Cariblatta   Blattellidae, Dictyoptera
Cariblatta lutea Saussure & Zehntner

Carnus   Carnidae, Diptera
Carnus hemapterus Nitzsch

Cataglyphis   Formicidae, Hymenoptera
Cataglyphis savignyi (Dufour)

Catallagia   Hystrichopsyllidae, Siphonaptera
Catallagia charlottensis (Baker)
Catallagia luski Schwan & Nelson
Catallagia mathesoni Jameson

Catharsius   Scarabaeidae, Coleoptera
Catharsius sagax Quensel
Catharsius tricornutus DeGeer

Cattasoma   Sarcophagidae, Diptera
Cattasoma mcalpinei de Souza Lopes

Cavernicola   Reduviidae, Hemiptera
Cavernicola lenti Barrett & Arias
Cavernicola pilosa Barber

Cebalges   Psoroptidae, Acari

Cebalgoides   Psoroptidae, Acari

Cediopsylla   Pulicidae, Siphonaptera
Cediopsylla simplex (Baker)

Celatoblatta   Blattidae, Dictyoptera
Celatoblatta brunni (Alfken)
Celatoblatta peninsularis Johns
Celatoblatta quinquemaculata Johns
Celatoblatta undulivitta (Walker)
Celatoblatta vulgaris Johns

Cenocorixa   Corixidae, Hemiptera

Centrophlebomyia *Centrophlebomyia anthropophaga* (Robineau–Desvoidy)	Piophilidae, Diptera
Centruroides	Buthidae, Scorpiones
Centruroides argentinus Werner	**see** *Centruroides margaritatus*
Centruroides danieli (Prado & Rios–Patiño)	**see** *Centruroides margaritatus*
Centruroides dasypus Mello–Leitão	**see** *Centruroides vittatus*
Centruroides exilicauda (Wood)	
Centruroides exsul (Meise)	
Centruroides gracilis (Latreille)	
Centruroides infamatus (Koch)	
Centruroides limpidus limpidus (Karsch)	
Centruroides limpidus tecomanus Hoffmann	
Centruroides margaritatus (Gervais)	
Centruroides nigrescens (Pocock)	
Centruroides noxius Hoffmann	
Centruroides pococki Sissom & Francke	
Centruroides sculpturatus Ewing	
Centruroides suffusus Pocock	
Centruroides testaceus (DeGeer)	
Centruroides vittatus (Say)	
Cephalonomia *Cephalonomia gallicola* (Ashmead)	Bethylidae, Hymenoptera
Cephalopina *Cephalopina titillator* (Clark)	Oestridae, Diptera
Cephenemyia	Oestridae, Diptera
Cephenemyia apicata Bennett & Sabrosky	
Cephenemyia auribarbis (Meigen)	
Cephenemyia jellisoni Townsend	
Cephenemyia phobifer (Clark)	
Cephenemyia stimulator (Clark)	
Cephenemyia trompe (Modeer)	
Cephenemyia ulrichii (Brauer)	
Ceratixodes	**see** *Ixodes*
Ceratixodes putus (Pickard–Cambridge)	**see** *Ixodes uriae*
Ceratoculicoides *Ceratoculicoides moravicus* Knoz	Ceratopogonidae, Diptera
Ceratophaga *Ceratophaga vastella* (Zeller)	Tineidae, Lepidoptera
Ceratophyllus	Ceratophyllidae, Siphonaptera
Ceratophyllus anisus Rothschild	
Ceratophyllus borealis Rothschild	
Ceratophyllus calcarifer Wagner	**see** *Megabothris calcarifer*
Ceratophyllus caspius Ioff & Argyropulo	**see** *Callopsylla caspia*
Ceratophyllus celsus Jordan	
Ceratophyllus chasteli Beaucournu, Monnat & Launay	
Ceratophyllus ciliatus ciliatus Baker	
Ceratophyllus ciliatus protinus Jordan	
Ceratophyllus columbae (Walckenaer & Gervais)	
Ceratophyllus consimilis Wagner	**see** *Nosopsyllus consimilis*
Ceratophyllus farreni Rothschild	
Ceratophyllus fasciatus (Bosc)	**see** *Nosopsyllus fasciatus*
Ceratophyllus forficus (Cai & Wu)	
Ceratophyllus gallinae (Schrank)	
Ceratophyllus garei Rothschild	
Ceratophyllus hamutus (Cai & Wu)	
Ceratophyllus hirundinis (Curtis)	

Ceratophyllus idius Jordan & Rothschild
Ceratophyllus indages Rothschild
Ceratophyllus iranus (Wagner & Argyropulo)   **see** *Nosopsyllus iranus*
Ceratophyllus laeviceps Wagner   **see** *Nosopsyllus laeviceps*
Ceratophyllus mokrzeckyi Wagner   **see** *Nosopsyllus mokrzeckyi*
Ceratophyllus niger Fox
Ceratophyllus paradoxus Scalon
Ceratophyllus penicilliger (Grube)   **see** *Amalaraeus penicilliger*
Ceratophyllus qinghaiensis Zhang & Ma
Ceratophyllus rusticus Wagner
Ceratophyllus sciurorum (Schrank)
Ceratophyllus spinosus Wagner
Ceratophyllus styx riparius Jordan & Rothschild
Ceratophyllus styx styx Rothschild
Ceratophyllus tesquorum Wagner   **see** *Citellophilus tesquorum*
Ceratophyllus turbidus Rothschild   **see** *Megabothris turbidus*
Ceratophyllus vagabundus insularis Rothschild
Ceratophyllus vagabundus vagabundus (Boheman)
Ceratophyllus vison Baker
Ceratophyllus walkeri Rothschild   **see** *Megabothris walkeri*

Ceratophyus   **see** *Typhaeus*

Ceratoppia   Metrioppiidae, Acari
Ceratoppia acuminata (Koch)

Cerceris   Sphecidae, Hymenoptera
Cerceris arenaria Linnaeus

Cerceustathia   Eustathiidae, Acari
Cerceustathia ruginosa Gaud

Cerodirphia   Saturniidae, Lepidoptera
Cerodirphia avenata araguensis Lemaire
Cerodirphia avenata avenata (Draudt)

Ceroptera   Sphaeroceridae, Diptera
Ceroptera sivinskii Marshall

Cervicola   **see** *Damalinia*

Chactas   Chactidae, Scorpiones
Chactas raymondhansi Francke & Boos

Chactopsis   Chactidae, Scorpiones

Chaetopsylla   Vermipsyllidae, Siphonaptera
Chaetopsylla floridensis (Fox)
Chaetopsylla globiceps (Taschenberg)
Chaetopsylla mirabilis Ioff & Argyropulo
Chaetopsylla rothschildi Kohaut
Chaetopsylla tuberculaticeps Bezzi

Chaetoravinia   Sarcophagidae, Diptera
Chaetoravinia advena (Walker)
Chaetoravinia almeidai Lopes

Chalybion   Sphecidae, Hymenoptera
Chalybion californicum (Saussure)

Chalybosoma   Tabanidae, Diptera
Chalybosoma cyaneoviridis (Macquart)   **see** *Cydistomyia cyanea*
Chalybosoma cyaneus (Wiedemann)   **see** *Cydistomyia cyanea*

Chaoborus
Chaoborus americanus (Johannsen)
Chaoborus anomalus Edwards
Chaoborus astictopus Dyar & Shannon
Chaoborus ceratopogones (Theobald)
Chaoborus crystallinus (DeGeer)
Chaoborus edulis Edwards
Chaoborus flavicans (Meigen)
Chaoborus fuscinervis Edwards
Chaoborus punctipennis (Say)

Chaoboridae, Diptera

Chauliacia
Chauliacia affinis Gaud

Eustathiidae, Acari

Chauliognathus
Chauliognathus pennsylvanicus (DeGeer)

Cantharidae, Coleoptera

Cheiracanthium
Cheiracanthium lawrencei Roewer
Cheiracanthium stratioticum L. Koch

Clubionidae, Araneae

Cheiridium

Cheiridiidae, Pseudoscorpiones

Cheiroseius
Cheiroseius curtipes (Halbert)

Ascidae, Acari

Chelacaropsis

Cheyletidae, Acari

Cheladonta
Cheladonta palawanensis Brown & Goff

Trombiculidae, Acari

Chelocnetha
Chelocnetha angustitarsis angustitarsis (Lundström)
Chelocnetha angustitarsis zaporojae Pavlichenko
Chelocnetha oresti Vorobets

see *Simulium*
see *Simulium angustitarse angustitarse*
see *Simulium angustitarse zaporojae*
see *Simulium oresti*

Chelopistes
Chelopistes meleagridis (Linnaeus)

Philopteridae, Phthiraptera

Cherax
Cherax tenuimanus (Smith)

Parastacidae, Decapoda

Cheyletia
Cheyletia papillifera Volgin

Cheyletidae, Acari

Cheyletiella
Cheyletiella blakei Smiley
Cheyletiella parasitivorax (Mégnin)
Cheyletiella yasguri Smiley

Cheyletiellidae, Acari

Cheyletus
Cheyletus aversor Rodendorf
Cheyletus fortis Oudemans
Cheyletus malaccensis Oudemans
Cheyletus tenuipilis Fain, Feldman-Muhsam & Mumcuoglu
Cheyletus trouessarti Oudemans

Cheyletidae, Acari

Chilacarus
Chilacarus martini Webb, Bennett & Loomis

Trombiculidae, Acari

Chimaerohelea
Chimaerohelea caligula Debenham

Ceratopogonidae, Diptera

Chinius *Chinius junlianensis* Leng	Psychodidae, Diptera
Chirobia *Chirobia cynopteri* Klompen, Lukoschus & Nadchatram	Teinocoptidae, Acari
Chirodiscoides *Chirodiscoides caviae* Hirst	Atopomelidae, Acari
Chirolaelaps *Chirolaelaps mystacinae* Heath, Bishop & Daniel	Laelapidae, Acari

Chironomus   Chironomidae, Diptera
Chironomus anthracinus Zetterstedt
Chironomus attenuatus Walker
Chironomus balatonicus Dévai, Wülker & Scholl
Chironomus barbatitarsis Kieffer   **see** *Kiefferulus barbatitarsis*
Chironomus californicus Johannsen   **see** *Dicrotendipes californicus*
Chironomus circumdatus (Kieffer)
Chironomus crassicaudatus Malloch
Chironomus decorus Johannsen
Chironomus flaviplumus Tokunaga
Chironomus frommeri Atchley & Martin
Chironomus macani Freeman
Chironomus maturus Johannsen
Chironomus pallidivittatus Malloch   **see** *Camptochironomus tentans*
Chironomus plumosus (Linnaeus)
Chironomus riparius piger Strenzke
Chironomus riparius riparius Meigen
Chironomus salinarius Kieffer
Chironomus staegeri Lundbeck
Chironomus tentans Fabricius   **see** *Camptochironomus tentans*
Chironomus thummi piger Strenzke   **see** *Chironomus riparius piger*
Chironomus thummi thummi (Kieffer)   **see** *Chironomus riparius riparius*
Chironomus vancouveri Michailova & Fischer
Chironomus yoshimatsui Martin & Sublette

Chiropterargas   **see** *Argas*

Chiroptonyssus   Macronyssidae, Acari
Chiroptonyssus haematophagus (Fonseca)

Chloromyia   Stratiomyiidae, Diptera
Chloromyia formosa (Scopoli)

Chlororhinia   Calliphoridae, Diptera

Chlorotabanus   Tabanidae, Diptera
Chlorotabanus inanis (Fabricius)
Chlorotabanus mexicanus (Linnaeus)

Chondracanthus   Chondracanthidae, Copepoda
Chondracanthus pinguis Wilson

Chorioptes   Psoroptidae, Acari
Chorioptes bovis (Hering)
Chorioptes caprae (Delafond & Bourguignon)   **see** *Chorioptes bovis*
Chorioptes equi (Gerlach)   **see** *Chorioptes bovis*
Chorioptes ovis (Zürn)   **see** *Chorioptes bovis*

Choristoneura   Tortricidae, Lepidoptera
Choristoneura fumiferana (Clemens)

ARTHROPODS OF MEDICAL AND VETERINARY IMPORTANCE

Chortoglyphus   Chortoglyphidae, Acari
Chortoglyphus sciuricola (Hyland & Fain)

Chrestomutilla   Mutillidae, Hymenoptera

Chrysomya   Calliphoridae, Diptera
Chrysomya albiceps (Wiedemann)
Chrysomya bezziana Villeneuve
Chrysomya chloropyga (Wiedemann)
Chrysomya chloropyga putoria (Wiedemann)   **see** Chrysomya putoria
Chrysomya marginalis (Wiedemann)   **see** Chrysomya regalis
Chrysomya megacephala (Fabricius)
Chrysomya nigripes Aubertin
Chrysomya pinguis (Walker)
Chrysomya putoria (Wiedemann)
Chrysomya regalis Robineau-Desvoidy
Chrysomya rufifacies (Macquart)
Chrysomya schoenigi Kurahashi & Magpayo
Chrysomya varipes (Macquart)
Chrysomya villeneuvii Patton
Chrysomya yayukae Kurahashi & Magpayo

Chrysops   Tabanidae, Diptera
Chrysops asbestos Philip
Chrysops ater Macquart
Chrysops atlanticus Pechuman
Chrysops atrinus Wang
Chrysops caecutiens (Linnaeus)
Chrysops centurionis Austen
Chrysops concavus Loew
Chrysops dawsoni Philip
Chrysops discalis Williston
Chrysops distinctipennis Austen
Chrysops excitans Walker
Chrysops fixissimus Walker
Chrysops flavidus Wiedemann
Chrysops fuliginosus Wiedemann
Chrysops fulvaster Osten Sacken
Chrysops furcatus Walker
Chrysops japonicus Wiedemann
Chrysops longicornis Macquart
Chrysops mitis Osten Sacken
Chrysops mlokosiewiczi Bigot
Chrysops nigripes Zetterstedt
Chrysops noctifer Osten Sacken
Chrysops obsoletus Wiedemann
Chrysops pictus Meigen   **see** Chrysops viduatus
Chrysops relictus Meigen
Chrysops sackeni Hine
Chrysops suavis Loew
Chrysops vanderwulpi kitaensis Hayakawa
Chrysops vanderwulpi vanderwulpi Kröber
Chrysops vanderwulpi yamatoensis Hayakawa
Chrysops venus Philip
Chrysops viduatus (Fabricius)
Chrysops zinzalus Philip

Chrysozona   **see** Haematopota

Chthamalus   Chthamalidae, Cirripedia
Chthamalus dalli Pilsbury
Chthamalus fissus Darwin

Ciconiphilus Ciconiphilus decimfasciatus (Boisduval & Lacordaire)	Menoponidae, Phthiraptera
Cimex Cimex dissimilis (Horváth) Cimex hemipterus (Fabricius) Cimex lectularius Linnaeus Cimex macrocephalus (Fieber) Cimex pipistrelli Jenyns	Cimicidae, Hemiptera see Cimex hemipterus
Circellium Circellium bacchus (Fabricius)	Scarabaeidae, Coleoptera
Citellophilus Citellophilus tesquorum (Wagner)	Ceratophyllidae, Siphonaptera
Cladotanytarsus Cladotanytarsus lewisi (Freeman)	Chironomidae, Diptera
Clavella Clavella adunca (Strøm)	Lernaeopodidae, Copepoda
Clavellisa Clavellisa cordata Wilson Clavellisa emarginata (Krøyer) Clavellisa pelione Tripathi Clavellisa scombri (Kurz) Clavellisa spinosa Wilson	Lernaeopodidae, Copepoda
Cleitosimulium Cleitosimulium argenteostriatum (Strobl)	Simuliidae, Diptera see Simulium argenteostriatum
Clitumnus Clitumnus extradentatus Brunner	see Baculum see Baculum extradentatum
Clogmia	Psychodidae, Diptera
Clunio Clunio marinus Haliday Clunio tsushimensis Tokunaga	Chironomidae, Diptera
Cnemidocoptes	see Knemidokoptes
Cnephia Cnephia dacotensis (Dyar & Shannon) Cnephia ornithophilia Davies, Peterson & Wood Cnephia pecuarum (Riley) Cnephia pilfreyi Davies & Györkös	Simuliidae, Diptera
Cnesia Cnesia dissimilis (Edwards)	Simuliidae, Diptera
Cnetha Cnetha keiseri Rubtsov Cnetha shogakii (Rubtsov) Cnetha subcostatum (Takahasi) Cnetha uchidai (Takahasi)	see Simulium see Simulium keiseri see Simulium shogakii see Simulium subcostatum see Simulium uchidai
Cobboldia Cobboldia elephantis (Cobbold)	Muscidae, Diptera
Cochliomyia Cochliomyia hominivorax (Coquerel) Cochliomyia macellaria (Fabricius)	Calliphoridae, Diptera

Coelopa
Coelopa frigida (Fabricius)
Coelopa vanduzeei Cresson

Coelopidae, Diptera

Coelotanypus
Coelotanypus scapularis (Loew)

Chironomidae, Diptera

Coilodes
Coilodes castanea Westwood

Scarabaeidae, Coleoptera

Colobathriglyphus
Colobathriglyphus malayensis Fain & Nadchatram

Acaridae, Acari

Colobomatus
Colobomatus kyphosus Sekerak

Philichthyidae, Copepoda

Colobopterus
Colobopterus fossor (Linnaeus)

Scarabaeidae, Coleoptera

Coloceras
Coloceras damicorne (Nitzsch)
Coloceras damicorne fahrenholzi Eichler
Coloceras piageti (Johnston & Harrison)

Goniodidae, Phthiraptera

see *Coloceras damicorne*

Colpa
Colpa interrupta (Fabricius)

Scoliidae, Hymenoptera

Colpocephalum
Colpocephalum meridionale Pérez–Jiménez *et al.*
Colpocephalum turbinatum Denny

Menoponidae, Phthiraptera

Columbicola
Columbicola claviformis (Denny)
Columbicola columbae bacillus (Giebel)
Columbicola columbae columbae (Linnaeus)

Philopteridae, Phthiraptera

Comperia
Comperia merceti (Compere)

Encyrtidae, Hymenoptera

Conicera
Conicera formosensis Brues

Phoridae, Diptera

Coorilla
Coorilla allisoni Mardon

Ischnopsyllidae, Siphonaptera

Copris
Copris anceus Olivier
Copris armatus Harold
Copris diversus Waterhouse
Copris elphenor Klug
Copris hispanus (Linnaeus)
Copris incertus Say
Copris laeviceps Harold
Copris lugubris Boheman
Copris lunaris (Linnaeus)
Copris pecuarius Lewis
Copris serius Nguyen–Phung
Copris umbilicatus Abeille de Perrin

Scarabaeidae, Coleoptera

Coproica
Coproica acutangula (Zetterstedt)
Coproica ferruginata (Stenhammer)
Coproica hirtula (Rondani)
Coproica vagans (Haliday)

Sphaeroceridae, Diptera

Copromyza *Copromyza fimetaria* (Meigen) *Copromyza frigipennis* (Spuler) *Copromyza pallifrons* Fallén *Copromyza similis* (Collin) *Copromyza singularis* (Spuler) *Copromyza vitripennis* Meigen	Sphaeroceridae, Diptera
Coprophanaeus *Coprophanaeus telamon corythus* *Coprophanaeus telamon telamon* Erichson	Scarabaeidae, Coleoptera
Coptopsylla *Coptopsylla africana* Wagner *Coptopsylla lamellifer* (Wagner) *Coptopsylla wassiliewi* (Wagner)	Coptopsyllidae, Siphonaptera
Coptotermes *Coptotermes acinaciformis* (Froggatt)	Rhinotermitidae, Isoptera
Coquillettidia *Coquillettidia buxtoni* (Edwards) *Coquillettidia chrysosoma* (Edwards) *Coquillettidia crassipes* (van der Wulp) *Coquillettidia fuscopennata* (Theobald) *Coquillettidia metallica* (Theobald) *Coquillettidia microannulata* (Theobald) *Coquillettidia nigricans* (Coquillett) *Coquillettidia perturbans* (Walker) *Coquillettidia richiardii* (Ficalbi) *Coquillettidia venezuelensis* (Theobald) *Coquillettidia xanthogaster* (Edwards)	Culicidae, Diptera
Cordylobia *Cordylobia anthropophaga* (Blanchard & Bérenger–Féra) *Cordylobia rodhaini* Gedoelst	Calliphoridae, Diptera
Corethrella *Corethrella appendiculata* Grabham *Corethrella brakeleyi* (Coquillett) *Corethrella metcalfi* McKeever *Corethrella wirthi* Stone	Chaoboridae, Diptera
Corrodopsylla *Corrodopsylla birulai* Ioff	Hystrichopsyllidae, Siphonaptera
Cosmiomma	Ixodidae, Acari
Cosmolaelaps *Cosmolaelaps gurabensis* Fox	**see** *Hypoaspis* **see** *Hypoaspis miles*
Cothonaspis	Eucoilidae, Hymenoptera
Craspedochaeta *Craspedochaeta punctipennis* (Wiedemann)	Anthomyiidae, Diptera
Craspedorrhynchus *Craspedorrhynchus aquilinus* Denny *Craspedorrhynchus fasciati* Gállego et al. *Craspedorrhynchus fraterculus* Eichler & Złotorzycka *Craspedorrhynchus melittoscopus* Nitzsch *Craspedorrhynchus nisi* Denny *Craspedorrhynchus pennati* Gállego et al. *Craspedorrhynchus platystomus* (Burmeister)	Philopteridae, Phthiraptera

Craspedorrhynchus rotundatus Piaget
Craspedorrhynchus spathulatus Giebel
Craspedorrhynchus subbuteonis Gállego et al.
Craspedorrhynchus triangularis Rudow

Crataerina   Hippoboscidae, Diptera
Crataerina hirundinis (Linnaeus)
Crataerina pallida (Latreille)

Creniola   Cymothoidae, Isopoda

Creophilus   Staphylinidae, Coleoptera
Creophilus erythrocephalus (Fabricius)
Creophilus maxillosus (Linnaeus)

Cricotopus   Chironomidae, Diptera
Cricotopus bicinctus (Meigen)
Cricotopus sylvestris (Fabricius)
Cricotopus tibialis Meigen
Cricotopus trifasciatus (Meigen)

Criniscansor   Myocoptidae, Acari
Criniscansor apodemi Fain, Mutinga & Lukoschus

Criokeron   Cheyletidae, Acari
Criokeron quintus (Domrow & Baker)
Criokeron thailandicus Fain & Lukoschus

Crivellia   **see** *Przhevalskiana*

Crocothemis   Libellulidae, Odonata
Crocothemis erythraea (Brullé)
Crocothemis servilia erythraea (Brullé)   **see** *Crocothemis erythraea*

Crotiscus   Trombiculidae, Acari
Crotiscus brennani Goff

Cryptocanthon   Scarabaeidae, Coleoptera
Cryptocanthon chiriquinus Howden & Gill
Cryptocanthon humidus Howden

Cryptocercus   Blattidae, Dictyoptera
Cryptocercus punctulatus Scudder

Cryptochironomus   Chironomidae, Diptera

Cryptocyclops   Cyclopidae, Copepoda

Cryptops   Cryptopidae, Chilopoda
Cryptops anomalans Newport

Cryptotylus   Tabanidae, Diptera
Cryptotylus unicolor (Wiedemann)

Ctenocephalides   Pulicidae, Siphonaptera
Ctenocephalides canis (Curtis)
Ctenocephalides felis damarensis Jordan
Ctenocephalides felis felis (Bouché)
Ctenocephalides felis orientis (Jordan)   **see** *Ctenocephalides orientis*
Ctenocephalides felis strongylus (Jordan)
Ctenocephalides orientis (Jordan)

Ctenocephalus   **see** *Ctenocephalides*
Ctenocephalus canis (Curtis)   **see** *Ctenocephalides canis*

Ctenoparia   Hystrichopsyllidae, Siphonaptera
Ctenoparia inopinata Rothschild
Ctenoparia jordani Smit
Ctenoparia propinqua Beaucournu & Gallardo
Ctenoparia topali Smit

Ctenophthalmus   Hystrichopsyllidae, Siphonaptera
Ctenophthalmus agyrtes agyrtes (Heller)
Ctenophthalmus agyrtes hanzaki Rosicky   **see** Ctenophthalmus agyrtes agyrtes
Ctenophthalmus agyrtes impavidus Jordan
Ctenophthalmus agyrtes peusianus Rosický
Ctenophthalmus andorrensis Smit
Ctenophthalmus apertus apertus Jordan & Rothschild
Ctenophthalmus apertus meylani Beaucournu et al.
Ctenophthalmus assimilis (Taschenberg)
Ctenophthalmus baeticus Rothschild
Ctenophthalmus bifidatus bifidatus Smit
Ctenophthalmus bifidatus mancus Beaucournu & Orhan
Ctenophthalmus bisoctodentatus Kolenati
Ctenophthalmus breviatus Wagner & Ioff
Ctenophthalmus calceatus cabirus Jordan & Rothschild
Ctenophthalmus calceatus calceatus Waterston
Ctenophthalmus dolichus Rothschild
Ctenophthalmus golovi Ioff & Tiflov
Ctenophthalmus harputus Aktoş
Ctenophthalmus lewisi Peus
Ctenophthalmus nobilis nobilis (Rothschild)
Ctenophthalmus nobilis vulgaris Smit
Ctenophthalmus orientalis (Wagner)
Ctenophthalmus pseudagyrtes Baker
Ctenophthalmus rettigi Rothschild
Ctenophthalmus russulae galloibericus Beaucournu & Lamaret
Ctenophthalmus russulae russulae Jordan & Rothschild
Ctenophthalmus teres Ioff & Argyropulo
Ctenophthalmus uncinatus (Wagner)
Ctenophthalmus wladimiri Isayeva-Gurvich

Ctenopsyllus   **see** Leptopsylla
Ctenopsyllus segnis Schönherr   **see** Leptopsylla segnis

Cuclotogaster   Philopteridae, Phthiraptera
Cuclotogaster cinereus (Nitzsch)
Cuclotogaster heterogrammicus (Nitzsch)
Cuclotogaster heterographa (Nitzsch)
Cuclotogaster obscurior Hopkins

Culex   Culicidae, Diptera
Culex adairi Kirkpatrick
Culex adamesi Sirivanakarn & Galindo
Culex aikenii (Aiken & Rowland)
Culex aikenii Dyar & Knab   **see** Culex quinquefasciatus
Culex albinervis Edwards
Culex alpha Séguy   **see** Culex theileri
Culex amazonensis (Lutz)
Culex andersoni andersoni Edwards
Culex andersoni abyssinicus Edwards
Culex annulirostris Skuse
Culex annulus Theobald
Culex antennatus (Becker)
Culex apicalis Adams
Culex australicus Dobrotworsky & Drummond
Culex azurini Toma, Miyagi & Cabrera
Culex bahamensis Dyar & Knab
Culex bidens Dyar

Culex bitaeniorhynchus Giles
Culex bonnei Dyar
Culex calloti Rioux & Pech see *Culex pipiens*
Culex castroi Casal & García
Culex caudelli (Dyar & Knab)
Culex cedecei Stone & Hair see *Culex taeniopus*
Culex cinereus Theobald
Culex coronator Dyar & Knab
Culex crybda Dyar
Culex decens Theobald
Culex delpontei Duret
Culex deserticola Kirkpatrick
Culex dispectus Bram
Culex dunni Dyar
Culex duttoni Theobald
Culex edwardsi Barraud
Culex epidesmus (Theobald)
Culex erectus Iglisch see *Culex pipiens*
Culex erraticus (Dyar & Knab)
Culex erythrothorax Dyar
Culex ethiopicus Edwards see *Culex bitaeniorhynchus*
Culex eumimetes Dyar & Knab see *Culex stigmatosoma*
Culex fatigans Wiedemann see *Culex quinquefasciatus*
Culex ferreri Duret
Culex fuscanus Wiedemann
Culex fuscocephalus Theobald
Culex gelidus Theobald
Culex glyptosalpinx Harbach, Peyton & Harrison
Culex guiarti Blanchard
Culex habilitator Dyar & Knab
Culex hainanensis Chen
Culex halifaxii Theobald
Culex harrisoni Sirivanakarn
Culex hayashii Yamada
Culex hortensis Ficalbi
Culex hutchinsoni Barraud
Culex idottus Dyar
Culex incidens Thomson see *Culiseta incidens*
Culex interfor Dyar
Culex intrincatus Brèthes
Culex invidiosus Theobald
Culex iyengari Mattingly & Rageau
Culex litwakae Harbach
Culex macrostylus Sirivanakarn & Ramalingam
Culex marquesensis Stone & Rosen
Culex martinii Medschid
Culex mathesoni Anduze see *Culex urichii*
Culex mauritanicus Callot see *Culex simpsoni*
Culex mimeticus Noé
Culex modestus Ficalbi
Culex molestus Forskål
Culex neavei Theobald
Culex nebulosus Theobald
Culex nigripalpus Theobald
Culex nigropunctatus Edwards
Culex ocellatus (Theobald)
Culex ocossa Dyar & Knab
Culex opisthopus Komp see *Culex taeniopus*
Culex oresbius Harbach & Rattanarithikul
Culex orientalis Edwards
Culex originator Gordon & Evans
Culex panocossa Dyar
Culex peccator Dyar & Knab
Culex pedroi Sirivanakarn & Belkin

Culex perexiguus Theobald
Culex perfuscus Edwards
Culex perplexus Leicester
Culex pervigilans Bergroth
Culex peus Speiser
Culex pilosus (Dyar & Knab)
Culex pipiens australicus Dobrotworsky & Drummond   **see** *Culex australicus*
Culex pipiens autogenicus Roubaud   **see** *Culex molestus*
Culex pipiens fatigans Wiedemann   **see** *Culex quinquefasciatus*
Culex pipiens molestus Forskål   **see** *Culex molestus*
Culex pipiens pallens Coquillett
Culex pipiens pipiens Linnaeus
Culex pipiens quinquefasciatus Say   **see** *Culex quinquefasciatus*
Culex plectoporpe Root
Culex poicilipes (Theobald)
Culex polynesiensis Marks
Culex portesi Senevet & Abonnenc
Culex pruina Theobald
Culex pseudovishnui Colless
Culex pusillus Macquart
Culex quasiguiarti Theobald
Culex quinquefasciatus Say
Culex restuans Theobald
Culex ribeirensis Forattini & Sallum
Culex richardgarciai Jeffery, Oothuman & Rudnick
Culex richei Klein
Culex roseni Belkin
Culex rubensis Sasa & Takahashi
Culex sacchettae Sirivanakarn & Jakob
Culex salinarius Coquillett
Culex saltanensis Dyar
Culex sangengluoensis Wang
Culex scanloni Bram
Culex serratimarge Root
Culex simpsoni Theobald
Culex sinaiticus Kirkpatrick
Culex sinensis Theobald
Culex siphanulatus Lourenço–de–Oliveira & Silva
Culex sitiens Wiedemann
Culex spathulatus Forattini & Sallum
Culex spissipes (Theobald)
Culex stigmatosoma Dyar
Culex taeniopus Dyar & Knab
Culex tarsalis Coquillett
Culex tenagius van Someren
Culex territans Walker
Culex thalassius Theobald
Culex theileri Theobald
Culex thriambus Dyar   **see** *Culex peus*
Culex tigripes Grandpré & Charmoy
Culex torrentium Martini
Culex torridus Iglisch   **see** *Culex pipiens*
Culex toviiensis Klein, Rivière & Séchan
Culex tritaeniorhynchus Giles
Culex tuberis Bohart
Culex univittatus Theobald
Culex urichii (Coquillett)
Culex vagans Wiedemann
Culex verutus Harbach
Culex vishnui Theobald
Culex weschei Edwards
Culex zombaensis Theobald

Culicoides   Ceratopogonidae, Diptera
Culicoides achrayi Kettle & Lawson
Culicoides actoni Smith
Culicoides albicans (Winnertz)
Culicoides alvarezi Ortiz
Culicoides anadyriensis Mirzaeva
Culicoides andicola Wirth & Lee
Culicoides arakawai (Arakawa)
Culicoides arboricola Root & Hoffman
Culicoides arschanicus Mirzaeva
Culicoides asiaticus Gutsevich & Smatov
Culicoides aterinervis Tokunaga
Culicoides austeni Carter, Ingram & Macfie
Culicoides australis Wirth & Jones   **see** *Culicoides variipennis australis*
Culicoides austropalpalis Lee & Reye
Culicoides bambusicola Lutz
Culicoides barbosai Wirth & Blanton
Culicoides belkini Wirth & Arnaud
Culicoides bermudensis Williams
Culicoides bernardae Itouta & Cornet
Culicoides biguttatus (Coquillett)
Culicoides brevipalpis Delfinado
Culicoides brevitarsis Kieffer
Culicoides brucei Austen
Culicoides cacticolus Wirth & Hubert
Culicoides cambodiensis Chu
Culicoides cataneii Clastrier
Culicoides cavaticus Wirth & Jones
Culicoides chaetophthalmus Amosova
Culicoides chiopterus (Meigen)
Culicoides circumscriptus Kieffer
Culicoides clastrieri Callot, Kremer & Deduit
Culicoides clavipalpis Mukerji
Culicoides clintoni Boorman
Culicoides coarctatus Clastrier & Wirth
Culicoides copiosus Root & Hoffman
Culicoides cordiger Macfie
Culicoides cornutus De Meillon
Culicoides crepuscularis Malloch
Culicoides cubitalis Edwards
Culicoides deanei Felippe-Bauer & Wirth
Culicoides debilipalpis Lutz
Culicoides denisoni Boorman
Culicoides denningi Foote & Pratt
Culicoides desertorum Gutsevich
Culicoides dewulfi Goetghebuer
Culicoides diabolicus Hoffman
Culicoides diamouanganai Itouta & Cornet
Culicoides distinctipennis Austen
Culicoides dzhafarovi Remm
Culicoides edeni Wirth & Blanton
Culicoides efferus Fox
Culicoides erairai Kôno & Takahasi
Culicoides eriodendroni Carter, Ingram & Macfie
Culicoides espinolai Felippe-Bauer *et al.*
Culicoides expallens Remm
Culicoides fascipennis (Staeger)
Culicoides filariferus Hoffman
Culicoides flavisomum Mirzaeva
Culicoides flavus Gornostaeva
Culicoides fluviatilis (Lutz)
Culicoides foxi Ortiz
Culicoides fulvithorax (Austen)
Culicoides furens (Poey)

Culicoides glabripennis Goetghebuer
Culicoides gluchovae Mirzaeva
Culicoides grahamii Austen
Culicoides grisescens Edwards
Culicoides grisescens flavus Gornostaeva   **see** *Culicoides flavus*
Culicoides gulbenkiani Caeiro
Culicoides guttipennis (Coquillett)
Culicoides guyanensis Floch & Abonnenc
Culicoides halophilus Kieffer
Culicoides henryi Lee & Reye
Culicoides heteroclitus Callot & Kremer
Culicoides hieroglyphicus Malloch
Culicoides hinmani Khalaf
Culicoides histrio Johannsen
Culicoides hollensis (Melander & Brues)
Culicoides huffi Causey
Culicoides imicola Kieffer
Culicoides impunctatus Goetghebuer
Culicoides indistinctus Khalaf
Culicoides insignis Lutz
Culicoides jamaicensis Edwards
Culicoides jamesi Fox
Culicoides kanagai Khamala & Kettle
Culicoides kibunensis Tokunaga
Culicoides kingi Austen
Culicoides knowltoni Beck
Culicoides kolymbiensis Boorman
Culicoides komarovi Mirzaeva
Culicoides lahillei Iches
Culicoides leopoldoi Ortiz
Culicoides leucostictus Kieffer
Culicoides limai Barretto
Culicoides longicollis Glukhova
Culicoides longior Hagan & Reye
Culicoides longipennis Khalaf
Culicoides loughnani Edwards
Culicoides machardyi Campbell & Pelham-Clinton   **see** *Culicoides manchuriensis*
Culicoides machardyi submaritimus Dzhafarov   **see** *Culicoides maritimus*
Culicoides maculatus Shiraki
Culicoides manchuriensis Tokunaga
Culicoides maritimus Kieffer
Culicoides maritimus submaritimus Dzhafarov   **see** *Culicoides maritimus*
Culicoides marksi Lee & Reye
Culicoides matsuzawai Tokunaga
Culicoides melleus (Coquillett)
Culicoides milnei Austen
Culicoides minasensis Felippe-Bauer
Culicoides mississippiensis Hoffman
Culicoides mohave Wirth
Culicoides musilator Kremer & Callot
Culicoides nanus Root & Hoffman
Culicoides neofagineus Wirth & Blanton
Culicoides newsteadi Austen
Culicoides niger Root & Hoffman
Culicoides nipponensis Tokunaga
Culicoides nivosus De Meillon
Culicoides notatus Delfinado
Culicoides nubeculosus (Meigen)
Culicoides obsoletus (Meigen)
Culicoides occidentalis Wirth & Jones   **see** *Culicoides variipennis occidentalis*
Culicoides odiatus Austen
Culicoides ohmorii Takahashi
Culicoides ornatus Taylor
Culicoides oxystoma Kieffer

Culicoides pallidicornis Kieffer
Culicoides pallidipennis Carter, Ingram & Macfie   **see** *Culicoides imicola*
Culicoides paradisionensis Boorman
Culicoides paraensis (Goeldi)
Culicoides paraflavescens Wirth & Hubert
Culicoides paucienfuscatus Barbosa
Culicoides peliliouensis Tokunaga
Culicoides peregrinus Kieffer
Culicoides phlebotomus (Williston)
Culicoides pictipennis (Staeger)
Culicoides pongsomiensis Chu
Culicoides praetermissus Carter, Ingram & Macfie   **see** *Culicoides leucostictus*
Culicoides pseudodiabolicus Fox
Culicoides pulicaris (Linnaeus)
Culicoides pulicaris punctatus (Meigen)   **see** *Culicoides punctatus*
Culicoides pumilus Winnertz
Culicoides punctatus (Meigen)
Culicoides puncticollis (Becker)
Culicoides pusillus Lutz
Culicoides pycnostictus Ingram & Macfie
Culicoides rangeli Ortiz & Mirsa
Culicoides reconditus Campbell & Pelham-Clinton
Culicoides riethi Kieffer
Culicoides riouxi Callot & Kremer
Culicoides rutshuruensis Goetghebuer
Culicoides ryckmani Wirth & Hubert
Culicoides saevus Kieffer
Culicoides salinarius Kieffer
Culicoides sanguisuga (Coquillett)
Culicoides schultzei (Enderlein)
Culicoides scoticus Downes & Kettle
Culicoides segnis Campbell & Pelham-Clinton
Culicoides sigaensis Tokunaga   **see** *Culicoides maculatus*
Culicoides similis Carter, Ingram & Macfie
Culicoides simulator Edwards
Culicoides sinanoensis Tokunaga
Culicoides sitiens Wirth & Hubert
Culicoides slovacus Országh
Culicoides sonorensis Wirth & Jones   **see** *Culicoides variipennis sonorensis*
Culicoides stellifer (Coquillett)
Culicoides stigmalis Wirth
Culicoides suarezi Rodriguez & Wirth
Culicoides subimmaculatus Lee & Reye
Culicoides sumatrae Macfie
Culicoides tavaresi Felippe-Bauer & Wirth
Culicoides tororoensis Khamala & Kettle
Culicoides toyamaruae Arnaud
Culicoides travassosi Forattini
Culicoides trinidadensis Hoffman
Culicoides trouilleti Itouta & Cornet
Culicoides truncorum Edwards
Culicoides turanicus Gutsevich & Smatov
Culicoides ustinovi Shevchenko
Culicoides variipennis australis Wirth & Jones
Culicoides variipennis occidentalis Wirth & Jones
Culicoides variipennis sonorensis Wirth & Jones
Culicoides variipennis variipennis (Coquillett)
Culicoides venustus Hoffman
Culicoides wadai Kitaoka
Culicoides zuluensis De Meillon

Culiseta   Culicidae, Diptera
Culiseta alaskaensis (Ludlow)
Culiseta annulata (Schrank)

Culiseta annulata subochrea (Edwards)   **see** Culiseta subochrea
Culiseta bergrothi (Edwards)
Culiseta fumipennis (Stephens)
Culiseta impatiens (Walker)
Culiseta incidens (Thomson)
Culiseta inconspicua (Lee)
Culiseta inornata (Williston)
Culiseta longiareolata (Macquart)
Culiseta melanura (Coquillett)
Culiseta minnesotae Barr
Culiseta morsitans (Theobald)
Culiseta nipponica LaCasse & Yamaguti
Culiseta ochroptera (Peus)
Culiseta particeps (Adams)
Culiseta subochrea (Edwards)

Cummingsiella   Philopteridae, Phthiraptera
Cummingsiella sellatus Burmeister

Cupiennius   Ctenidae, Araneae
Cupiennius salei (Keyserling)

Cuterebra   Cuterebridae, Diptera
Cuterebra americana (Fabricius)
Cuterebra buccata (Fabricius)
Cuterebra emasculator Fitch
Cuterebra fontinella fontinella Clark
Cuterebra fontinella grisea Coquillett
Cuterebra grisea Coquillett   **see** Cuterebra fontinella grisea
Cuterebra horripilum Clark
Cuterebra lepivora Coquillett
Cuterebra lepusculi Townsend   **see** Cuterebra princeps
Cuterebra princeps (Austen)
Cuterebra ruficrus (Austen)

Cyaneolytta   Meloidae, Coleoptera
Cyaneolytta depressicornis (Laporte)
Cyaneolytta granulipennis (Laporte)
Cyaneolytta iridescens (Haag-Rutenberg)
Cyaneolytta signifrons (Fåhraeus)

Cybister   Dytiscidae, Coleoptera
Cybister japonicus Sharp

Cyclopodia   Nycteribiidae, Diptera
Cyclopodia sykesii (Westwood)

Cyclops   Cyclopidae, Copepoda
Cyclops bicuspidatus bicuspidatus Claus   **see** Diacyclops bicuspidatus bicuspidatus
Cyclops bicuspidatus thomasi Forbes   **see** Diacyclops thomasi
Cyclops leuckarti Claus   **see** Mesocyclops leuckarti
Cyclops vernalis Fischer   **see** Acanthocyclops vernalis

Cydistomyia   Tabanidae, Diptera
Cydistomyia cyanea (Wiedemann)

Cymothoa   Cymothoidae, Isopoda

Cynomya   Calliphoridae, Diptera
Cynomya mortuorum (Linnaeus)

Cynomyopsis   Calliphoridae, Diptera
Cynomyopsis cadaverina (Robineau-Desvoidy)

Cyphogenia *Cyphogenia gibba* (Fischer von Waldheim)	Tenebrionidae, Coleoptera
Cyphosaccus	Peltogastridae, Cirripedia
Cyrtolaelaps *Cyrtolaelaps berlesei* Chelebiev *Cyrtolaelaps chiropterae* Karg *Cyrtolaelaps mucronatus* (G. & R. Canestrini)	Rhodacaridae, Acari
Cytodites *Cytodites nudus* (Vizioli)	Cytoditidae, Acari

D

Daidalotarsonemus *Daidalotarsonemus hewitti* Mahunka	Tarsonemidae, Acari
Daira *Daira perlata* (Herbst)	Xanthidae, Decapoda

Damalinia *Damalinia antidorcus* (Bedford) *Damalinia bovis* (Linnaeus) *Damalinia breviceps* (Rudow) *Damalinia caprae* (Gurlt) *Damalinia cornuta* (Gervais) *Damalinia equi* (Denny) *Damalinia limbata* (Gervais) *Damalinia lipeuroides* (Mégnin) *Damalinia ovis* (Schrank) *Damalinia parallela* (Osborn) *Damalinia pelea* Bedford *Damalinia reduncae* (Bedford)	Trichodectidae, Phthiraptera **see** *Bovicola bovis* **see** *Bovicola caprae* **see** *Werneckiella equi* **see** *Bovicola limbata* **see** *Bovicola ovis*

Dasyhelea *Dasyhelea dufouri* (Laboulbène) *Dasyhelea leptocladus* Remm *Dasyhelea mutabilis* (Coquillett)	Ceratopogonidae, Diptera
Dasyphora *Dasyphora pratorum* (Meigen)	Muscidae, Diptera
Dasyponyssus *Dasyponyssus neivai* Fonseca	Dasyponyssidae, Acari
Dasypsyllus *Dasypsyllus gallinulae* (Dale)	Ceratophyllidae, Siphonaptera
Dasyrhamphis *Dasyrhamphis algirus* (Macquart) *Dasyrhamphis anthracinus* (Meigen) *Dasyrhamphis denticornis* (Enderlein)	Tabanidae, Diptera
Decticus *Decticus verrucivorus* (Linnaeus)	Tettigoniidae, Orthoptera
Degeeriella *Degeeriella fulva* (Giebel)	Philopteridae, Phthiraptera
Deinocerites *Deinocerites cancer* Theobald *Deinocerites magnus* (Theobald) *Deinocerites pseudes* Dyar & Knab	Culicidae, Diptera

Delotelis
Delotelis telegoni Rothschild

Hystrichopsyllidae, Siphonaptera

Deltochilum
Deltochilum lobipes Bates
Deltochilum orbignyi Blanchard
Deltochilum scabrisculum Bates

Scarabaeidae, Coleoptera

Demodex
Demodex acutipes Bukvar & Preisler
Demodex antechini Nutting & Sweatman
Demodex aries Desch
Demodex aurati Nutting
Demodex bantengi Firda, Nutting & Sweatman
Demodex bovis Stiles
Demodex brevis Akbulatova
Demodex buccalis Bukva
Demodex cafferi Nutting & Guilfoy
Demodex canis Leydig
Demodex caprae Railliet
Demodex cati Mégnin
Demodex cervi (Prietsch)
Demodex cervi (Vanselow)
Demodex criceti Nutting & Rauch
Demodex equi Railliet
Demodex flagellurus Bukva
Demodex folliculorum (Simon)
Demodex folliculorum hominis Leydig
Demodex foveolator Bukva
Demodex ghanensis Oppong, Lee & Yasin
Demodex kutzeri Bukva
Demodex nanus Hirst
Demodex neoopisthosomae Desch *et al.*
Demodex ovis Railliet
Demodex peromysci Lombert, Lukoschus & Whitaker
Demodex phylloides Csokor
Demodex rosus Bukva
Demodex sabani Desch, Lukoschus & Nadchatram
Demodex spelaea Desch, Lukoschus & Nadchatram
Demodex suis Railliet
Demodex tauri Bukva

Demodicidae, Acari

see *Demodex kutzeri*

see *Demodex folliculorum*

see *Demodex phylloides*

Dendrolimus
Dendrolimus punctatus (Walker)

Lasiocampidae, Lepidoptera

Dendrophaonia
Dendrophaonia querceti (Bouché)

Anthomyiidae, Diptera

Dendrophilus
Dendrophilus xavieri Marseul

Histeridae, Coleoptera

Dentocarpus
Dentocarpus macrotrichus Dusbábek & de la Cruz
Dentocarpus silvai eumopsicola Fain
Dentocarpus silvai silvai Dusbábek & de la Cruz

Chirodiscidae, Acari

Deois
Deois flavopicta (Stål)

Cercopidae, Hemiptera

Dermacarus
Dermacarus hylandi Fain
Dermacarus newyorkensis Fain

Glycyphagidae, Acari

see *Glycyphagus newyorkensis*

Dermacentonomma

Ixodidae, Acari

Dermacentor	Ixodidae, Acari
Dermacentor albipictus (Packard)	
Dermacentor andersoni Stiles	
Dermacentor atrosignatus Neumann	
Dermacentor auratus Supino	
Dermacentor compactus Neumann	
Dermacentor daghestanicus Olenev	
Dermacentor marginatus (Sulzer)	
Dermacentor montanus Filippova & Panova	
Dermacentor nitens Neumann	**see** *Anocentor nitens*
Dermacentor niveus Neumann	
Dermacentor nuttalli Olenev	
Dermacentor occidentalis Marx	
Dermacentor parumapertus Neumann	
Dermacentor pavlovskyi Olenev	
Dermacentor pictus (Hermann)	
Dermacentor raskemensis Serdyukova	
Dermacentor reticulatus (Fabricius)	
Dermacentor rhinocerinus (Denny)	
Dermacentor silvarum Olenev	
Dermacentor steini Schulze	
Dermacentor taiwanensis Sugimoto	
Dermacentor ushakovae Filippova & Panova	
Dermacentor variabilis (Say)	
Dermadelema	Trombiculidae, Acari
Dermadelema furmani (Gould)	
Dermadelema lynnae Pomeroy & Loomis	
Dermadelema mojavense Pomeroy & Loomis	
Dermadelema sleeperi Pomeroy & Loomis	
Dermanyssus	Dermanyssidae, Acari
Dermanyssus gallinae (DeGeer)	
Dermanyssus hirundinis (Hermann)	
Dermanyssus passerinus Berlese & Trouessart	
Dermatobia	Cuterebridae, Diptera
Dermatobia hominis (Linnaeus jr.)	
Dermatophagoides	Pyroglyphidae, Acari
Dermatophagoides chelidonis (Hull)	**see** *Hirstia chelidonis*
Dermatophagoides culinae De Leon	**see** *Dermatophagoides farinae*
Dermatophagoides deanei Galvão & Guitton	**see** *Dermatophagoides neotropicalis*
Dermatophagoides evansi Fain, Hughes & Johnston	
Dermatophagoides farinae Hughes	
Dermatophagoides microceras Griffiths & Cunnington	
Dermatophagoides neotropicalis Fain & van Bronswijk	
Dermatophagoides pteronyssinus (Trouessart)	
Dermatophagoides scheremetewskyi Bogdanov	
Dermestes	Dermestidae, Coleoptera
Dermestes ater DeGeer	
Dermestes ater domesticus Germar	**see** *Dermestes ater*
Dermestes caninus Germar	
Dermestes frischii Kugelann	
Dermestes lardarius Linnaeus	
Dermestes maculatus DeGeer	
Dermestes peruvianus Laporte	
Dermestes shaanxiensis Cao	
Dermestes tessellatocollis Motschulsky	
Dermoergasilus	Ergasilidae, Copepoda
Dermoergasilus acanthopagri Byrnes	
Dermoergasilus mugilis Oldewage & van As	

Dermoglyphus *Dermoglyphus passerinus* Gaud	Dermoglyphidae, Acari
Desmometopa *Desmometopa m-nigrum* Zetterstedt	Milichiidae, Diptera
Dexopollenia *Dexopollenia nigra* Kurahashi	Calliphoridae, Diptera
Diabrotica *Diabrotica balteata* Le Conte	Chrysomelidae, Coleoptera
Diacyclops *Diacyclops bicuspidatus bicuspidatus* (Claus) *Diacyclops thomasi* (Forbes)	Cyclopidae, Copepoda
Dialytes	Scarabaeidae, Coleoptera
Diamanus *Diamanus montanus* (Baker)	**see** *Oropsylla* **see** *Oropsylla montana*
Diapria *Diapria conica* (Fabricius)	Diapriidae, Hymenoptera
Diaptomus *Diaptomus leptopus* Forbes	Diaptomidae, Copepoda **see** *Aglaodiaptomus leptopus*
Dibrachys *Dibrachys cavus* (Walker)	Pteromalidae, Hymenoptera
Dichetophora *Dichetophora obliterata* (Fabricius)	Sciomyzidae, Diptera
Dichotomius *Dichotomius carolinus* (Linnaeus) *Dichotomius centralis* (Harold) *Dichotomius yucatanus* (Bates)	Scarabaeidae, Coleoptera
Dicranogonus *Dicranogonus ancoripalpus* Atyeo & Gaud *Dicranogonus botryodes* Atyeo & Gaud *Dicranogonus melanopis* Atyeo & Gaud *Dicranogonus phimosi* Atyeo & Gaud *Dicranogonus sculpturatus* Atyeo & Gaud *Dicranogonus theristici* Atyeo & Gaud	Pterolichidae, Acari
Dicrotendipes *Dicrotendipes californicus* (Johannsen) *Dicrotendipes fusconotatus* (Kieffer)	Chironomidae, Diptera
Didactylia *Didactylia tenuipunctata* Bordat	Scarabaeidae, Coleoptera
Didymocentrus *Didymocentrus lesueurii* (Gervais) *Didymocentrus minor* Francke *Didymocentrus waeringi* Francke	Diplocentridae, Scorpiones
Dinopsyllus *Dinopsyllus lypusus* Jordan & Rothschild	Hystrichopsyllidae, Siphonaptera
Dipetalogaster *Dipetalogaster maxima* (Uhler)	Reduviidae, Hemiptera

Diplacodes	Libellulidae, Odonata
Diplacodes lefebvrii (Rambur)	
Diplacodes trivialis (Rambur)	
Diplocentrus	Diplocentridae, Scorpiones
Diplocentrus diablo Stockwell & Nilsson	
Diplonychus	Belostomatidae, Hemiptera
Diplonychus indicus Venkatesan & Rao	
Diplonychus nepoides (Fabricius)	
Diplonychus rusticus (Fabricius)	
Diploptera	Blaberidae, Dictyoptera
Diploptera punctata (Eschscholtz)	
Dipolaelaps	Laelapidae, Acari
Dipolaelaps chimmarogalis Gu	
Dipolaelaps jiangkouensis Gu	
Dipolaelaps longisetosus Huang	
Dipolaelaps nepalensis Till	
Dipolaelaps soriculi Huang	
Dipolaelaps taibaiensis Huang	
Dirhinus	Chalcididae, Hymenoptera
Dirhinus himalayanus Westwood	
Dirhinus pachycerus Masi	
Discalida	Pseudomopidae, Dictyoptera
Discalida pallidimarginia Woo, Guo & Li	
Discomyza	Ephydridae, Diptera
Discomyza maculipennis (Wiedemann)	
Distigmesikya	Pterolichidae, Acari
Distigmesikya deroptyi Faccini & Atyeo	
Distosimulium	**see** *Prosimulium*
Distosimulium daisetsense (Uemoto, Okazawa & Onishi)	**see** *Prosimulium daisetsense*
Docophthirus	Polyplacidae, Phthiraptera
Docophthirus acinetus Waterston	
Dohrniphora	Phoridae, Diptera
Dohrniphora cornuta (Bigot)	
Dolichoderus	Formicidae, Hymenoptera
Dolichovespula	Vespidae, Hymenoptera
Dolichovespula arctica Rohwer	
Dolichovespula arenaria (Fabricius)	
Dolichovespula maculata (Linnaeus)	
Dolichovespula norvegica (Fabricius)	
Dolichovespula norvegicoides (Sladen)	
Dolichovespula saxonica (Fabricius)	
Dolichovespula sylvestris (Scopoli)	
Doloisia	Trombiculidae, Acari
Doloisia iranensis Goff	
Dolomedes	Dolomedidae, Araneae
Dolomedes fimbriatus (Clerck)	
Doratopsylla	Hystrichopsyllidae, Siphonaptera
Doratopsylla birulai Ioff	**see** *Corrodopsylla birulai*

Doratopsylla blarinae Fox
Doratopsylla dasycnema cuspis Rothschild
Doratopsylla dasycnema dasycnema (Rothschild)

Dorcadia   Vermipsyllidae, Siphonaptera
Dorcadia xijiensis Zhang & Dang

Dorypteryx   Psyllipsocidae, Psocoptera
Dorypteryx domestica (Smithers)

Drepanocerus   Scarabaeidae, Coleoptera

Drino   Tachinidae, Diptera
Drino imberbis (Wiedemann)   **see** *Palexorista imberbis*

Drosophila   Drosophilidae, Diptera
Drosophila auraria Peng
Drosophila funebris (Fabricius)
Drosophila melanogaster Meigen
Drosophila persimilis Dobzhansky & Epling
Drosophila pseudoobscura Frolowa
Drosophila repleta Wollaston
Drosophila simulans Sturtevant

Drymaplaneta   Blattidae, Dictyoptera
Drymaplaneta variegata (Shelford)

Drymeia   Muscidae, Diptera
Drymeia vicana (Harris)

Dryomyza   Dryomyzidae, Diptera
Dryomyza anilis Fallén

Dubininia   Xolalgidae, Acari
Dubininia melopsittaci Atyeo & Gaud

Dugesiella   Theraphosidae, Araneae
Dugesiella californica (Ausserer)   **see** *Avicularia californicum*
Dugesiella echina Chamberlin   **see** *Rhechostica echina*

Dytiscus   Dytiscidae, Coleoptera
Dytiscus dauricus Gebler
Dytiscus verticalis Say

E
Echidnophaga   Pulicidae, Siphonaptera
Echidnophaga gallinacea (Westwood)
Echidnophaga murina (Tiraboschi)

Echinolaelaps   **see** *Laelaps*
Echinolaelaps echidnina (Berlese)   **see** *Laelaps echidnina*

Echinonyssus   **see** *Hirstionyssus*
Echinonyssus apoensis (Delfinado)   **see** *Hirstionyssus apoensis*
Echinonyssus blanchardi (Trouessart)   **see** *Hirstionyssus blanchardi*
Echinonyssus butantanensis (Fonseca)   **see** *Hirstionyssus butantanensis*
Echinonyssus confucianus (Hirst)   **see** *Hirstionyssus confucianus*
Echinonyssus cynomys (Radford)   **see** *Hirstionyssus cynomys*
Echinonyssus hilli (Jameson)   **see** *Hirstionyssus hilli*
Echinonyssus isabellinus (Oudemans)   **see** *Hirstionyssus isabellinus*
Echinonyssus talpae (Zemskaya)   **see** *Hirstionyssus talpae*

Echinophthirius   Echinophthiriidae, Phthiraptera
Echinophthirius horridus (von Olfers)

Ectatomma *Ectatomma tuberculatum* (Olivier)	Formicidae, Hymenoptera
Ectinorus *Ectinorus levipes* Jordan & Rothschild	Rhopalopsyllidae, Siphonaptera
Ectobius *Ectobius lapponicus* (Linnaeus) *Ectobius sylvestris* (Poda)	Pseudomopidae, Dictyoptera
Ectocyclops *Ectocyclops rubescens* Brady	Cyclopidae, Copepoda
Ectopsocopsis *Ectopsocopsis cryptomeriae* (Enderlein)	Peripsocidae, Psocoptera
Ectopsocus *Ectopsocus briggsi* MacLachlan	Ectopsocidae, Psocoptera
Edocla *Edocla slateri* Distant	Reduviidae, Hemiptera
Efferia *Efferia kondratieffi* Bullington & Lavigne	Asilidae, Diptera
Elachisoma *Elachisoma afrotropicum* Papp *Elachisoma bajzae* Papp *Elachisoma braacki* Papp	Sphaeroceridae, Diptera
Elasmopus *Elasmopus bampo* Barnard	Melitidae, Amphipoda
Elasmus *Elasmus polistis* Burks	Elasmidae, Hymenoptera
Elgiva *Elgiva connexa* (Steyskal) *Elgiva divisa* (Loew) *Elgiva elegans* Orth & Knutson *Elgiva pacnowesa* Orth & Knutson *Elgiva solicita* (Harris)	Sciomyzidae, Diptera
Elliptorhina *Elliptorhina brunneri* (Butler)	Blattidae, Dictyoptera
Eltonella	**see** *Ascoschoengastia*
Enallagma *Enallagma carunculatum* Morse *Enallagma civile* (Hagen)	Coenagriidae, Odonata
Encarsia *Encarsia flava* Shafee *Encarsia transvena* Timberlake	Aphelinidae, Hymenoptera **see** *Encarsia transvena*
Encelimyia *Encelimyia aurigena* de Souza Lopes *Encelimyia pelenguensis* de Souza Lopes	Sarcophagidae, Diptera
Encentridophorus *Encentridophorus similis* Cook	Unionicolidae, Acari

Enderleinellus *Enderleinellus longiceps* Kellogg & Ferris *Enderleinellus nitzschi* Fahrenholz	Enderleinellidae, Phthiraptera
Endochironomus *Endochironomus nigricans* Johannsen	Chironomidae, Diptera
Engelemyia *Engelemyia inops* (Walker)	Sarcophagidae, Diptera
Enithares *Enithares templetoni* (Kirby)	Notonectidae, Hemiptera
Entepherus *Entepherus laminipes* Bere	Cecropidae, Copepoda
Entonyssus *Entonyssus halli* Ewing	Entonyssidae, Acari
Eomenacanthus *Eomenacanthus stramineus* (Nitzsch)	**see** *Menacanthus* **see** *Menacanthus stramineus*

Epicauta — Meloidae, Coleoptera
Epicauta callosa LeConte
Epicauta conferta (Say)
Epicauta fabricii (LeConte)
Epicauta ficta Werner — **see** *Epicauta obesa*
Epicauta funebris Horn
Epicauta immaculata Say
Epicauta lemniscata (Fabricius) — **see** *Epicauta occidentalis*
Epicauta maculata Say
Epicauta murina (LeConte)
Epicauta obesa (Chevrolat)
Epicauta occidentalis Werner
Epicauta pennsylvanica (DeGeer)
Epicauta pestifera Werner
Epicauta sericans LeConte
Epicauta torsa (LeConte)
Epicauta vittata (Fabricius)

Epidermoptes *Epidermoptes bilobatus* (Rivolta)	Epidermoptidae, Acari
Epilampra *Epilampra maya* Rehn	Epilampridae, Dictyoptera
Epilineutes *Epilineutes globosus* (O. Pickard-Cambridge)	Theridiosomatidae, Araneae
Epimyodex *Epimyodex crocidurae* Fain, Lukoschus & Rosmalen *Epimyodex microti* Fain, Lukoschus & Rosmalen *Epimyodex talpae* Fain & Orts	Cloacaridae, Acari
Epipenaeon *Epipenaeon elegans* Chopra	Bopyridae, Isopoda
Epitedia *Epitedia faceta* (Rothschild) *Epitedia wenmanni* (Rothschild)	Hystrichopsyllidae, Siphonaptera
Eratyrus *Eratyrus mucronatus* Stål	Reduviidae, Hemiptera

ARTHROPODS OF MEDICAL AND VETERINARY IMPORTANCE

Erebuniella	Tabanidae, Diptera
Erebuniella cordiger (Meigen)	**see** *Tabanus cordiger*
Eretmapodites	Culicidae, Diptera
Eretmapodites chrysogaster Graham	
Eretmapodites hamoni Grjebine	
Eretmapodites quinquevittatus Theobald	
Ergasilus	Ergasilidae, Copepoda
Ergasilus borneoensis Yamaguti	
Ergasilus celestis Mueller	**see** *Ergasilus versicolor*
Ergasilus cerastes Roberts	
Ergasilus ceylonensis Fernando & Hanek	
Ergasilus ilani Oldewage & van As	
Ergasilus labracis Krøyer	
Ergasilus lizae Krøyer	
Ergasilus malnadensis Venkatershappa *et al.*	
Ergasilus mirabilis Oldewage & van As	
Ergasilus nanus Beneden	
Ergasilus sieboldi Nordmann	
Ergasilus versicolor Wilson	
Eriphia	Xanthidae, Decapoda
Eriphia sebana (Shaw & Nodlee)	
Eristalis	Syrphidae, Diptera
Eristalis tenax (Linnaeus)	
Erithrodiplax	Libellulidae, Odonata
Erithrodiplax umbrata (Linnaeus)	
Euboettcheria	Sarcophagidae, Diptera
Euboettcheria subducta (Lopes)	
Euborellia	Anisolabiidae, Dermaptera
Euborellia annulipes (Lucas)	
Eubrachiella	Lernaeopodidae, Copepoda
Eubrachiella antarctica (Quidor)	
Eubrachylaelaps	Laelapidae, Acari
Eubrachylaelaps rotundus Fonseca	
Eubranchipus	Branchipodidae, Branchiopoda
Eubranchipus bundyi Forbes	
Eucampsipoda	Nycteribiidae, Diptera
Eucampsipoda hyrtlii (Kolenati)	
Euchaeta	Euchaetidae, Copepoda
Euchaeta elongata Esterly	
Euchaeta norvegica Boeck	
Euchromius	Pyralidae, Lepidoptera
Euchromius ocelleus Haworth	
Eucoila	Eucoilidae, Hymenoptera
Eucoila hunteri Crawford	**see** *Ganaspidium hunteri*
Eucyclops	Cyclopidae, Copepoda
Eudiaptomus	Diaptomidae, Copepoda
Eudiaptomus gracilis (Sars)	

Eudusbabekia *Eudusbabekia rosickyi* (Dusbábek) *Eudusbabekia samsinaki* (Dusbábek)	Myobiidae, Acari
Eugamasus *Eugamasus berlesei* Willmann *Eugamasus lunulatus* (Müller)	Parasitidae, Acari
Euhoplopsyllus *Euhoplopsyllus glacialis affinis* (Baker) *Euhoplopsyllus glacialis glacialis* (Taschenberg)	Pulicidae, Siphonaptera
Eukiefferiella *Eukiefferiella minor* (Edwards)	Chironomidae, Diptera
Eulaelaps *Eulaelaps dremomydis* Gu & Wang *Eulaelaps heptacanthus* Yang & Gu *Eulaelaps huzhuensis* Yang & Gu *Eulaelaps linggangis* Wen & Yan *Eulaelaps novus* Vitzthum *Eulaelaps oudemansi* Turk *Eulaelaps stabularis* (Koch)	Laelapidae, Acari
Eulinognathus *Eulinognathus allactagae* Johnson *Eulinognathus americanus* Ewing *Eulinognathus biuncatus* Ferris *Eulinognathus bolivianus* Werneck *Eulinognathus euchoreutei* Cais *Eulinognathus patagonicus* Castro & Cicchino *Eulinognathus torquatus* Castro *Eulinognathus wernecki* Castro & Cicchino	Polyplacidae, Phthiraptera
Eumenes	Eumenidae, Hymenoptera
Eumolpianus *Eumolpianus eumolpi* (Rothschild)	Ceratophyllidae, Siphonaptera
Euodynerus *Euodynerus foraminatus* (Saussure)	Eumenidae, Hymenoptera
Euoniticellus *Euoniticellus africanus* (Harold) *Euoniticellus fulvus* (Goeze) *Euoniticellus intermedius* (Reiche) *Euoniticellus pallipes* (Fabricius)	**see** *Oniticellus* **see** *Oniticellus africanus* **see** *Oniticellus fulvus* **see** *Oniticellus intermedius* **see** *Oniticellus pallipes*
Euonthophagus *Euonthophagus bedeli* (Reitter) *Euonthophagus bedeli conterminus* (Petrovitz) *Euonthophagus conterminus* (Petrovitz) *Euonthophagus rapillyi* Baraud	**see** *Onthophagus* **see** *Onthophagus bedeli* **see** *Onthophagus conterminus* **see** *Onthophagus conterminus* **see** *Onthophagus conterminus*
Euparaphyto *Euparaphyto peruana* Tibana & Souza Lopes	Sarcophagidae, Diptera
Eupelops *Eupelops occultus* (Koch)	Phenopelopidae, Acari
Eupolyphaga *Eupolyphaga sinensis* (Walker)	Blattidae, Dictyoptera

Euproctis *Euproctis chrysorrhoea* (Linnaeus) *Euproctis edwardsi* (Newman)	Lymantriidae, Lepidoptera
Eupteromalus	**see** *Trichomalopsis*
Euroglyphus *Euroglyphus longior* (Trouessart) *Euroglyphus maynei* (Cooreman)	Pyroglyphidae, Acari **see** *Gymnoglyphus longior*
Eurycotis *Eurycotis floridana* (Walker)	Blattidae, Dictyoptera
Eurymella	**see** *Megachile*
Euryparasitus *Euryparasitus emarginatus* (Koch)	Rhodacaridae, Acari
Eurypelma *Eurypelma californicum* (Ausserer)	Theraphosidae, Araneae **see** *Avicularia californicum*
Euschoengastia *Euschoengastia furmani* Gould *Euschoengastia indica* (Hirst) *Euschoengastia jamesoni* (Brennan) *Euschoengastia ohioensis* Farrell *Euschoengastia rotundata* (Shluger) *Euschoengastia whitakeri* Wrenn	Trombiculidae, Acari **see** *Dermadelema furmani* **see** *Ascoschoengastia indica*
Euscorpius *Euscorpius flavicaudis* (DeGeer) *Euscorpius italicus* (Herbst)	Chactidae, Scorpiones
Eusimulium *Eusimulium angustipes* (Edwards) *Eusimulium angustitarse* (Lundström) *Eusimulium aureum* (Fries) *Eusimulium bicorne* (Dorogostaĭskiĭ *et al.*) *Eusimulium canonicolum* Dyar & Shannon *Eusimulium carpathicum* Knoz *Eusimulium costatum* (Friederichs) *Eusimulium cryophilum* Rubtsov *Eusimulium emarginatum* (Davies, Peterson & Wood) *Eusimulium erimoense* Ono *Eusimulium euryadminiculum* (Davies) *Eusimulium fallisi* Golini *Eusimulium latipes* (Meigen) *Eusimulium latizonum* Rubtsov *Eusimulium rendalense* Golini *Eusimulium securiforme* Rubtsov *Eusimulium subcostatum* Takahasi *Eusimulium uchidai* Takahasi *Eusimulium usovae* Golini *Eusimulium vernum* (Macquart)	**see** *Simulium* **see** *Simulium angustipes* **see** *Simulium angustitarse* **see** *Simulium aureum* **see** *Simulium bicorne* **see** *Simulium canonicolum* **see** *Simulium carpathicum* **see** *Simulium costatum* **see** *Simulium cryophilum* **see** *Simulium emarginatum* **see** *Simulium erimoense* **see** *Simulium euryadminiculum* **see** *Simulium fallisi* **see** *Simulium latipes* **see** *Simulium latizonum* **see** *Simulium rendalense* **see** *Simulium securiforme* **see** *Simulium subcostatum* **see** *Simulium uchidai* **see** *Simulium usovae* **see** *Simulium vernum*
Eustathia	Eustathiidae, Acari
Euthycera *Euthycera cribrata* (Rondani)	Sciomyzidae, Diptera
Eutrombicula *Eutrombicula alfreddugesi* (Oudemans) *Eutrombicula autumnalis* (Shaw) *Eutrombicula batatas* (Linnaeus)	Trombiculidae, Acari **see** *Neotrombicula autumnalis*

Eutrombicula belkini (Gould)
Eutrombicula cinnabaris (Ewing)   **see** Eutrombicula alfreddugesi
Eutrombicula macropus (Womersley)
Eutrombicula multisetosa (Ewing)
Eutrombicula splendens (Ewing)

Evania   Evaniidae, Hymenoptera
Evania appendigaster (Linnaeus)

Ewingana   Myobiidae, Acari
Ewingana doreyae Dusbábek

Exhyalanthrax   Bombyliidae, Diptera
Exhyalanthrax alliopterus (Hesse)
Exhyalanthrax transiens (Bezzi)

Exopalpiger   Ixodidae, Acari
Exopalpiger trianguliceps (Birulya)   **see** Ixodes trianguliceps

Eyndhovenia   Spinturnicidae, Acari
Eyndhovenia brachypus Sun, Wang & Wang
Eyndhovenia euryalis euryalis (G. Canestrini)
Eyndhovenia euryalis oudemansi (Eyndhoven)

F
Fainalges   Xolalgidae, Acari
Fainalges apicosetiger Mejía-Gonzalez & Pérez
Fainalges brevissimus Mejía-Gonzalez & Pérez
Fainalges longissimus Mejía-Gonzalez & Pérez

Falculifer   Falculiferidae, Acari
Falculifer rostratus (Buchholz)

Fannia   Fanniidae, Diptera
Fannia barbata Stein
Fannia canicularis (Linnaeus)
Fannia femoralis (Stein)
Fannia heydenii (Wiedemann)
Fannia howardi Malloch
Fannia leucosticta (Meigen)
Fannia obscurinervis (Stein)
Fannia pusio (Wiedemann)
Fannia scalaris (Fabricius)
Fannia vesparia (Meade)

Fanthamia   Ceratopogonidae, Diptera
Fanthamia cardinis De Meillon & Wirth
Fanthamia ornatipennis (De Meillon)

Farhangia   Pygiopsyllidae, Siphonaptera

Farrelioides   **see** Euschoengastia
Farrelioides jamesoni (Brennan)   **see** Euschoengastia jamesoni

Felicola   Trichodectidae, Phthiraptera
Felicola subrostratus (Burmeister)

Felistrophorus   Listrophoridae, Acari
Felistrophorus radovskyi (Tenorio)

Feronia   **see** Pterstichus

Ficalbia *Ficalbia minima* (Theobald) *Ficalbia uniformis* (Theobald)	Culicidae, Diptera
Figites	Cynipidae, Hymenoptera
Filistata *Filistata hibernalis* Hentz	Filistatidae, Araneae
Finlaya *Finlaya khasiana* Barraud	**see** *Aedes* **see** *Aedes formosensis*
Fleuria *Fleuria lacustris* Kieffer	Chironomidae, Diptera
Fonsecalges	Psoroptidae, Acari
Forcipomyia *Forcipomyia anabaenae* Chan & Saunders *Forcipomyia phlebotomoides* Bangerter *Forcipomyia stimulans* (de Meijere) *Forcipomyia townsvillensis* (Taylor) *Forcipomyia velox* (Winnertz)	Ceratopogonidae, Diptera
Forelius *Forelius foetidus* (Buckley)	Formicidae, Hymenoptera
Forficula *Forficula auricularia* Linnaeus	Forficulidae, Dermaptera
Formica *Formica clara* Forel *Formica cunicularia* Latreille *Formica podzolica* Francoeur *Formica polyctena* Förster *Formica pratensis* Retzius *Formica rufa* Linnaeus *Formica sanguinea* Latreille *Formica subnuda* Emery	Formicidae, Hymenoptera
Freyana *Freyana anatina* (Koch)	Freyanidae, Acari
Frontopsylla *Frontopsylla elatoides elatoides* Wagner *Frontopsylla elatoides longa* Mikulin *Frontopsylla frontalis frontalis* (Rothschild) *Frontopsylla frontalis postcurva* Liu, Wu & Wu *Frontopsylla rotunditruncata* Cai & Liu	Leptopsyllidae, Siphonaptera
Furipterobia	Myobiidae, Acari
Fuscuropoda *Fuscuropoda vegetans* (DeGeer)	Uropodidae, Acari

G

Gabronthus *Gabronthus mgogoricus* Tottenham *Gabronthus thermarum* (Aubé)	Staphylinidae, Coleoptera
Gahrliepia *Gahrliepia disparunguis disparunguis* (Oudemans) *Gahrliepia disparunguis pingue* (Gater) *Gahrliepia ligula* (Radford)	Trombiculidae, Acari **see** *Schoengastiella ligula*

Gahrliepia longipedalis Yu & Yang
Gahrliepia lorentzi Goff
Gahrliepia parapacifica Chen, Hsu & Wang
Gahrliepia saduski Womersley
Gahrliepia sagitta Goff
Gahrliepia serrata Brown & Goff
Gahrliepia sinensis Goff

Gallilichus   Syringobiidae, Acari
Gallilichus hiregoudari D'Souza & Jagannath

Galumna   Galumnidae, Acari
Galumna flabellifera Hammer
Galumna niliaca Al-Assiuty *et al.*

Gamasellodes   Ascidae, Acari
Gamasellodes vermivorax Walter

Gamasholaspis   Parholaspidae, Acari
Gamasholaspis eothenomydis Gu

Gammarus   Gammaridae, Amphipoda
Gammarus fasciatus Say
Gammarus lacustris Sars
Gammarus pseudolimnaeus Bousfield
Gammarus pulex (Linnaeus)
Gammarus roeseli Gervais

Ganaspidium   Eucoilidae, Hymenoptera
Ganaspidium hunteri Crawford

Gasterophilus   Gasterophilidae, Diptera
Gasterophilus equi (Clark)   **see** *Gasterophilus intestinalis*
Gasterophilus flavipes (Olivier)   **see** *Gasterophilus haemorrhoidalis*
Gasterophilus haemorrhoidalis (Linnaeus)
Gasterophilus inermis (Brauer)
Gasterophilus intestinalis (DeGeer)
Gasterophilus meridionalis (Pillers & Evans)
Gasterophilus nasalis (Linnaeus)
Gasterophilus nigricornis (Loew)
Gasterophilus ovis (Linnaeus)   **see** *Oestrus ovis*
Gasterophilus pecorum (Fabricius)
Gasterophilus ternicinctus Gedoelst
Gasterophilus veterinus (Clark)   **see** *Gasterophilus nasalis*

Geckobia   Pterygosomatidae, Acari
Geckobia andoharonomaitsoensis Haitlinger
Geckobia bataviensis Vitzthum
Geckobia cosymboti Cuy   **see** *Geckobia bataviensis*
Geckobia gleadoviana Hirst   **see** *Geckobia bataviensis*
Geckobia ifanadianaensis Haitlinger
Geckobia mananjaryensis Haitlinger
Geckobia nepalii Hiregauder   **see** *Geckobia bataviensis*
Geckobia samambavyensis Haitlinger

Geckobiella   Pterygosomatidae, Acari
Geckobiella texana (Banks)

Gedoelstia   Oestridae, Diptera
Gedoelstia haessleri Gedoelst

Geolycosa   Lycosidae, Araneae
Geolycosa godeffroyi (L. Koch)   **see** *Lycosa godeffroyi*

Geomydoecus — Trichodectidae, Phthiraptera
Geomydoecus davidhafneri Price & Hellenthal
Geomydoecus neotruncatus Hellenthal & Price
Geomydoecus quadridentatus Price & Emerson
Geomydoecus telli Price & Hellenthal
Geomydoecus truncatus Werneck

Geomylichus — Listrophoridae, Acari
Geomylichus californicus Fain, Whitaker & Thomas
Geomylichus deserti Fain & Whitaker
Geomylichus dipodomius (Radford)
Geomylichus formosus Fain & Whitaker
Geomylichus mexicanus Fain
Geomylichus microdipodops Fain & Whitaker
Geomylichus multistriatus Fain, Whitaker & Thomas
Geomylichus postscutatus Fain
Geomylichus texanus Fain, Whitaker, Schwan & Lukoschus
Geomylichus utahensis Fain & Whitaker

Geotrupes — Geotrupidae, Coleoptera
Geotrupes albarracinus (Wagner)
Geotrupes auratus Motschulsky
Geotrupes cavicollis Bates
Geotrupes intermedius (Costa)
Geotrupes laevigatus cobosi Baraud
Geotrupes laevigatus laevigatus (Fabricius)
Geotrupes pyrenaeus (Charpentier)
Geotrupes sericeus (Jekel)
Geotrupes spiniger (Marsham)
Geotrupes stercorarius (Linnaeus)
Geotrupes stercorosus (Scriba)
Geotrupes vernalis (Linnaeus)

Gerris — Gerridae, Hemiptera
Gerris lacustris (Linnaeus)

Geusibia — Leptopsyllidae, Siphonaptera
Geusibia longihilla Zhang & Liu
Geusibia minutiprominula minutiprominula Zhang & Liu
Geusibia minutiprominula ningshanensis Zhang & Liu

Gigantodax — Simuliidae, Diptera
Gigantodax chilense (Philippi)

Gigantolaelaps — Laelapidae, Acari
Gigantolaelaps amazonae Furman
Gigantolaelaps goyanensis Fonseca
Gigantolaelaps guimaraesi Lizaso
Gigantolaelaps mattogrossensis (Fonseca)
Gigantolaelaps oudemansi Fonseca
Gigantolaelaps peruviana (Ewing)
Gigantolaelaps vitzthumi Fonseca
Gigantolaelaps wolffsohni (Oudemans)

Gigliolella — Armilliferidae, Pentastomida
Gigliolella brumpti (Giglioli)

Glaciopsyllus — Ceratophyllidae, Siphonaptera
Glaciopsyllus antarcticus Smit & Dunnet

Gliricola — Gyropidae, Phthiraptera
Gliricola porcelli (Schrank)

Glossina   Glossinidae, Diptera
Glossina austeni Newstead
Glossina brevipalpis Newstead
Glossina caliginea Austen
Glossina frezili Gouteux
Glossina fusca congolensis Newstead & Evans
Glossina fusca fusca (Walker)
Glossina fuscipes fuscipes Newstead
Glossina fuscipes quanzensis Pires
Glossina longipalpis Wiedemann
Glossina longipennis Corti
Glossina medicorum Austen
Glossina morsitans centralis Machado
Glossina morsitans morsitans Westwood
Glossina morsitans submorsitans Newstead
Glossina nigrofusca Newstead
Glossina pallicera Bigot
Glossina pallidipes Austen
Glossina palpalis gambiensis Vanderplank
Glossina palpalis palpalis (Robineau-Desvoidy)
Glossina swynnertoni Austen
Glossina tabaniformis Westwood
Glossina tachinoides Westwood

Glossophagocarpus   Chirodiscidae, Acari
Glossophagocarpus cubanus de la Cruz

Glycyphagus   Glycyphagidae, Acari
Glycyphagus cadaverum (Schrank)   see Lepidoglyphus destructor
Glycyphagus destructor (Schrank)   see Lepidoglyphus destructor
Glycyphagus domesticus (DeGeer)
Glycyphagus hypudaei (Koch)
Glycyphagus newyorkensis (Fain)
Glycyphagus privatus Oudemans
Glycyphagus zapus Fain, Spicka, Jones & Whitaker

Glyptholaspis   Macrochelidae, Acari
Glyptholaspis americana (Berlese)
Glyptholaspis confusa (Foà)

Glyptotendipes   Chironomidae, Diptera
Glyptotendipes barbipes (Staeger)
Glyptotendipes gripekoveni (Kieffer)
Glyptotendipes pallens (Meigen)
Glyptotendipes paripes (Edwards)
Glyptotendipes polytomus (Kieffer)   see Glyptotendipes pallens
Glyptotendipes tokunagai Sasa

Gnathia   Gnathiidae, Isopoda
Gnathia calva Vanhöffen

Gnatocerus   Tenebrionidae, Coleoptera
Gnatocerus cornutus (Fabricius)

Gnus   see Simulium
Gnus cholodkovskii (Rubtsov)   see Simulium cholodkovskii
Gnus daisense (Takahasi)   see Simulium daisense

Gohieria   Glycyphagidae, Acari
Gohieria fusca (Oudemans)

Gomphostilbia   see Simulium
Gomphostilbia shogakii (Rubtsov)   see Simulium shogakii

Gonarcticus Gonarcticus arcticus (Becker)	Scathophagidae, Diptera
Goniocotes Goniocotes bidentatus (Scopoli) Goniocotes chrysocephalus Giebel Goniocotes gallinae (DeGeer) Goniocotes hologaster (Nitzsch) Goniocotes microthorax (Stephens) Goniocotes rectangulatus (Nitzsch)	Goniodidae, Phthiraptera **see** Campanulotes bidentatus **see** Goniocotes gallinae
Goniodes Goniodes astrocephalus (Burmeister) Goniodes australis Emerson & Price Goniodes colchici Denny Goniodes dispar Burmeister Goniodes dissimilis Denny Goniodes gigas (Taschenberg) Goniodes leipoae Emerson & Price Goniodes numidae Mjöberg Goniodes pavonis (Linnaeus) Goniodes tetraophasis Chou & Liu	Goniodidae, Phthiraptera
Gonolabis Gonolabis marginalis (Dohrn)	Anisolabiidae, Dermaptera
Grassomyia Grassomyia dreyfussi (Parrot) Grassomyia squamipleuris (Newstead)	Psychodidae, Diptera
Greniera Greniera dobyi Beaucournu–Saguez & Braverman Greniera nairica Terterian	Simuliidae, Diptera
Gromphadorhina Gromphadorhina brunneri Butler Gromphadorhina portentosa (Schaum)	Oxyhaloidae, Dictyoptera **see** Elliptorhina brunneri
Gryllodes Gryllodes sigillatus (Walker)	Gryllidae, Orthoptera
Gryllus Gryllus bimaculatus DeGeer Gryllus desertus Pallas Gryllus sigillatus (Walker)	Gryllidae, Orthoptera **see** Melanogryllus desertus **see** Gryllodes sigillatus
Gryphopsylla Gryphopsylla hetera Lewis & Jones	Pygiopsyllidae, Siphonaptera
Guatemalichus Guatemalichus tachornis Cruz, Cuervo & Dusbábek	Pyroglyphidae, Acari
Guntheria Guntheria forbesi Goff Guntheria platalea Domrow Guntheria shareli Domrow	Trombiculidae, Acari
Gymnodia Gymnodia cilifera (Malloch) Gymnodia lasiopa van Emden	Muscidae, Diptera
Gymnoglyphus Gymnoglyphus longior (Trouessart)	Pyroglyphidae, Acari

Gymnopais	Simuliidae, Diptera
Gymnopleurus *Gymnopleurus aciculatus* (Gebler) *Gymnopleurus mopsus* (Pallas) *Gymnopleurus sinuatus* (Olivier)	Scarabaeidae, Coleoptera
Gynaecoplotrupes	Scarabaeidae, Coleoptera
Gynaikothrips *Gynaikothrips ficorum* (Marchal)	Phlaeothripidae, Thysanoptera
Gyrostigma *Gyrostigma pavesii* (Corti) *Gyrostigma rhinocerontis* (Hope)	Oestridae, Diptera **see** *Gyrostigma rhinocerontis*

H

Habrobracon	Braconidae, Hymenoptera
Hadrurus *Hadrurus arizonensis* Ewing	Iuridae, Scorpiones
Haemadipsus *Haemadipsus leporis* Blagoveshchenskiĭ *Haemadipsus lyriocephalus* (Burmeister) *Haemadipsus setoni* Ewing	**see** *Haemodipsus* **see** *Haemodipsus leporis* **see** *Haemodipsus lyriocephalus* **see** *Haemodipsus setoni*
Haemagogus *Haemagogus albomaculatus* Theobald *Haemagogus anastasionis* Dyar *Haemagogus capricornii* Lutz *Haemagogus celeste* Dyar & Nuñez Tovar *Haemagogus equinus* Theobald *Haemagogus falco* Kumm *et al.* *Haemagogus janthinomys* Dyar *Haemagogus leucocelaenus* (Dyar & Shannon)	Culicidae, Diptera **see** *Haemagogus janthinomys*
Haemaphysalis *Haemaphysalis aciculifer* Warburton *Haemaphysalis anomaloceraea* Teng & Cui *Haemaphysalis bancrofti* Warburton & Nuttall *Haemaphysalis bispinosa* Neumann *Haemaphysalis bremneri* Roberts *Haemaphysalis campanulata* Warburton *Haemaphysalis concinna* Koch *Haemaphysalis cooleyi* Bedford *Haemaphysalis cornigera* Neumann *Haemaphysalis cornigera shimoga* Trapido & Hoogstraal *Haemaphysalis cornupunctata* Hoogstraal & Varma *Haemaphysalis cretica* Senevet & Caminopetros *Haemaphysalis cuspidata* Warburton *Haemaphysalis demidovae* Emel'yanova *Haemaphysalis flava* Neumann *Haemaphysalis hispanica* Gil Collado *Haemaphysalis hoodi* Warburton & Nuttall *Haemaphysalis houyi* Nuttall & Warburton *Haemaphysalis humerosa* Warburton & Nuttall *Haemaphysalis hyracophila* Hoogstraal, Walker & Neitz *Haemaphysalis hystricis* Supino *Haemaphysalis inermis* Birulya *Haemaphysalis intermedia* Warburton & Nuttall *Haemaphysalis japonica* Warburton *Haemaphysalis kinneari* Warburton *Haemaphysalis kohlsi* Aragão & Fonseca	Ixodidae, Acari **see** *Haemaphysalis shimoga* **see** *Haemaphysalis sulcata*

Haemaphysalis kopetdaghica Kerbabaev
Haemaphysalis kutchensis Hoogstraal & Trapido
Haemaphysalis kyasanurensis Trapido *et al.*
Haemaphysalis leachii (Audouin)
Haemaphysalis leporispalustris (Packard)
Haemaphysalis longicornis Neumann
Haemaphysalis neumanni Dönitz   **see** *Haemaphysalis longicornis*
Haemaphysalis norvali Hoogstraal & Wassef
Haemaphysalis otophila Schulze   **see** *Haemaphysalis parva* Neumann, 1897
Haemaphysalis papuana Thorell
Haemaphysalis papuana kinneari Warburton   **see** *Haemaphysalis kinneari*
Haemaphysalis paraleachi Camicas *et al.*
Haemaphysalis parva Neumann, 1897
Haemaphysalis parva Neumann, 1908   **see** *Haemaphysalis intermedia*
Haemaphysalis pedetes Hoogstraal
Haemaphysalis pospelovashtromae Hoogstraal
Haemaphysalis primitiva Teng
Haemaphysalis punctata Canestrini & Fanzago
Haemaphysalis qinghaiensis Teng
Haemaphysalis shimoga Trapido & Hoogstraal
Haemaphysalis silacea Robinson
Haemaphysalis spinigera Neumann
Haemaphysalis spinulosa Neumann
Haemaphysalis sulcata Canestrini & Fanzago
Haemaphysalis turturis Nuttall & Warburton
Haemaphysalis verticalis Itagaki, Noda & Yamaguchi

Haematobia   Muscidae, Diptera
Haematobia atripalpis Bezzi   **see** *Haematobosca atripalpis*
Haematobia crassipalpis Ringdahl   **see** *Haematobosca stimulans*
Haematobia exigua de Meijere   **see** *Haematobia irritans exigua*
Haematobia irritans exigua de Meijere
Haematobia irritans irritans (Linnaeus)
Haematobia minuta (Bezzi)
Haematobia sanguisugens Austen   **see** *Haematobosca stimulans*
Haematobia stimulans (Meigen)   **see** *Haematobosca stimulans*
Haematobia thirouxi potans (Bezzi)
Haematobia thirouxi thirouxi (Roubaud)
Haematobia thirouxi titillans (Bezzi)   **see** *Haematobia titillans*
Haematobia titillans (Bezzi)

Haematobosca   Muscidae, Diptera
Haematobosca atripalpis (Bezzi)
Haematobosca crassipalpis (Ringdahl)   **see** *Haematobosca stimulans*
Haematobosca stimulans (Meigen)

Haematomyzus   Haematomyzidae, Phthiraptera
Haematomyzus elephantis Piaget
Haematomyzus porci Emerson & Price

Haematopinus   Haematopinidae, Phthiraptera
Haematopinus asini (Linnaeus)
Haematopinus equi Simmonds   **see** *Haematopinus macrocephalus*
Haematopinus eurysternus Denny
Haematopinus macrocephalus (Burmeister)
Haematopinus quadripertusus Fahrenholz
Haematopinus suis (Linnaeus)
Haematopinus tuberculatus (Burmeister)

Haematopota   Tabanidae, Diptera
Haematopota americana Osten Sacken
Haematopota bealesi Coher
Haematopota bullatifrons Austen
Haematopota champlaini (Philip)

Haematopota coronata Austen
Haematopota crassicornis Wahlberg
Haematopota cynthiae Coher
Haematopota dissimilis Ricardo
Haematopota eugeniae Portillo & Schacht
Haematopota excipula Coher
Haematopota gobindai Coher
Haematopota italica Meigen
Haematopota lacessens Austen
Haematopota lambi Villeneuve
Haematopota meridionalis (Strobl)   see *Chrysops caecutiens*
Haematopota ocelligera (Kröber)
Haematopota pallens Loew
Haematopota pandazisi (Kröber)
Haematopota pluvialis pluvialis (Linnaeus)
Haematopota pluvialis tristis Bigot
Haematopota punctulata Macquart
Haematopota quadrifenestrata Burger
Haematopota scutellata (Olsuf'ev, Moucha & Chvála)
Haematopota subcylindrica Pandelle
Haematopota sumelae Timmer
Haematopota tamerlani Szilady
Haematopota tristis Bigot   see *Haematopota pluvialis tristis*
Haematopota turkestanica (Kröber)
Haematopota vimoli Coher

Haematosiphon   Cimicidae, Hemiptera
Haematosiphon inodorus (Dugèes)

Haemobaphes   Pennellidae, Copepoda
Haemobaphes diceraus Wilson
Haemobaphes intermedius Kabata

Haemodipsus   Polyplacidae, Phthiraptera
Haemodipsus leporis Blagoveshchenskiĭ
Haemodipsus lyriocephalus (Burmeister)
Haemodipsus setoni Ewing

Haemogamasus   Laelapidae, Acari
Haemogamasus ambulans (Thorell)
Haemogamasus bregetovae Mrciak
Haemogamasus dauricus Bregetova
Haemogamasus hirsutus Berlese
Haemogamasus horridus Michael
Haemogamasus mandschuricus Vitzthum
Haemogamasus nidi Michael
Haemogamasus nidiformes Bregetova
Haemogamasus qinghaiensis Yang & Gu
Haemogamasus reidi Ewing
Haemogamasus serdjukovae (Bregetova)
Haemogamasus trapezoideus Teng & Pan

Haemolaelaps   see *Androlaelaps*
Haemolaelaps casalis (Berlese)   see *Androlaelaps casalis*
Haemolaelaps glasgowi (Ewing)   see *Androlaelaps fahrenholzi*

Haffneria   Philopteridae, Phthiraptera
Haffneria grandis (Piaget)

Halictus   Apidae, Hymenoptera

Haliplus   Haliplidae, Coleoptera
Haliplus immaculicollis Harris

Halolaelaps	Halolaelapidae, Acari
Harpagophalla	Sarcophagidae, Diptera
Harpagophalla kempi Senior White	
Harpagoxenus	Formicidae, Hymenoptera
Harpagoxenus sublaevis (Nylander)	
Hatschekia	Hatschekiidae, Copepoda
Heaslipia	Trombiculidae, Acari
Heaslipia gateri (Womersley & Heaslip)	
Hebardina	Blattidae, Dictyoptera
Hebardina agaboides Gerstaecker	
Hebecnema	Muscidae, Diptera
Hebecnema umbratica (Meigen)	
Hecamede	Ephydridae, Diptera
Hecamede albicans (Meigen)	
Hectopsylla	Pulicidae, Siphonaptera
Hectopsylla psittaci Frauenfeld	
Heizmannia	Culicidae, Diptera
Heizmannia discrepans (Edwards)	
Helenicula	Trombiculidae, Acari
Helenicula abaensis Wang, Zhai & Chen	
Helicophagella	**see** *Sarcophaga*
Helicophagella melanura (Meigen)	**see** *Sarcophaga melanura*
Helina	Muscidae, Diptera
Helina dibrachiata Fang, Li & Deng	
Helina inflata Fang, Li & Deng	
Helina leptinocorpus Fang & Fan	
Helina xizangensis Fang & Fan	
Heliocis	Oedemeridae, Coleoptera
Heliocis repanda (Horn)	
Heliographa	Muscidae, Diptera
Heliographa acumicornis Shinonaga & Pont	
Heliographa albistriata Shinonaga & Pont	
Heliographa aurantiaca Stein	
Heliographa bicolorata Shinonaga & Pont	
Heliographa bismarckensis Shinonaga & Pont	
Heliographa excellens Stein	
Heliographa fasciata Stein	**see** *Heliographa steini*
Heliographa gressitti Shinonaga & Pont	
Heliographa insignis Stein	
Heliographa steini Shinonaga & Pont	
Helodon	**see** *Prosimulium*
Helodon multicaulis Popov	**see** *Prosimulium multicaulis*
Hemianax	Aeshnidae, Odonata
Hemianax papuensis (Burmeister)	
Hemifreyana	Avenzoariidae, Acari
Hemifreyana kurbanovae Mironov	

Hemilucilia *Hemilucilia flavifacies* (Engel) *Hemilucilia hermanlenti* Pinto de Mello *Hemilucilia semidiaphana* Rondani	Calliphoridae, Diptera **see** *Hemilucilia semidiaphana*
Hemimerus *Hemimerus bouvieri* Chopard	Hemimeridae, Dermaptera
Hemioniscus *Hemioniscus balani* Buchholz	Cryptoniscidae, Isopoda
Hemipepsis *Hemipepsis ustulata* Dahlbom	Pompilidae, Hymenoptera
Hemipimpla *Hemipimpla pulchripennis* (Saussure)	**see** *Camptotypus* **see** *Camptotypus pulchripennis*
Hemipyrellia *Hemipyrellia bougainvillia* Kurahashi *Hemipyrellia fernandica* (Macquart) *Hemipyrellia ligurriens* (Wiedemann)	Calliphoridae, Diptera
Henschoutedenia *Henschoutedenia flexivitta* (Walker)	Oxyhaloidae, Dictyoptera
Heptatoma *Heptatoma pellucens* (Fabricius)	Tabanidae, Diptera
Heptaulacus *Heptaulacus villosus* (Gyllenhal)	Scarabaeidae, Coleoptera
Hermetia *Hermetia illucens* (Linnaeus) *Hermetia remittens* Walker	Stratiomyidae, Diptera
Herpetacarus *Herpetacarus tiantai* Hsu & Hsu	Trombiculidae, Acari
Herpyllus *Herpyllus loricatus* (L. Koch)	Gnaphosidae, Araneae **see** *Scotophaeus loricatus*
Hershkovitzia *Hershkovitzia cabala* Peterson & Lacey	Nycteribiidae, Diptera
Hesperocimex *Hesperocimex coloradensis* List	Cimicidae, Hemiptera
Hesperoperla *Hesperoperla pacifica* (Banks)	Perlidae, Plecoptera
Heterodoxus *Heterodoxus ampullatus* Kéler *Heterodoxus spiniger* (Enderlein)	Boopiidae, Phthiraptera
Heterognathia	Gnathiidae, Isopoda
Heterometrus *Heterometrus bengalensis* (Koch) *Heterometrus fulvipes* (Koch) *Heterometrus swammerdami* Simon	Scorpionidae, Scorpiones
Heteronebo *Heteronebo muchmorei* Francke & Sissom *Heteronebo vachoni* Francke	Diplocentridae, Scorpiones **see** *Heteronebo vachoni*

Heteropezina *Heteropezina bharatika* Kashyap & Grover	Cecidomyiidae, Diptera
Heteropoda *Heteropoda venatoria* (Linnaeus)	Heteropodidae, Araneae
Heterosilpha *Heterosilpha aenescens* (Casey)	Silphidae, Coleoptera
Heterothops *Heterothops stiglundbergi* Israelson	Staphylinidae, Coleoptera
Hiereoblatta *Hiereoblatta cassidea* Eschscholtz	Polyphagidae, Dictyoptera
Hieroglyphus *Hieroglyphus banian* (Fabricius)	Acrididae, Orthoptera
Hippelates *Hippelates bilineatus* de Meijere *Hippelates bishoppi* Sabrosky *Hippelates collusor* (Townsend) *Hippelates pallipes* (Loew) *Hippelates pusio* Loew	Chloropidae, Diptera **see** *Cadrema pallida* var. *bilineata* **see** *Liohippelates bishoppi* **see** *Liohippelates collusor* **see** *Liohippelates pallipes* **see** *Liohippelates pusio*
Hippobosca *Hippobosca camelina* Leach *Hippobosca equina* Linnaeus *Hippobosca longipennis* Fabricius *Hippobosca maculata* Leach *Hippobosca rufipes* von Olfers *Hippobosca variegata* Megerle	Hippoboscidae, Diptera **see** *Hippobosca variegata*
Hippocentrodes *Hippocentrodes desmotes* Philip	Tabanidae, Diptera
Hippodamia *Hippodamia convergens* (Guérin-Méneville)	Coccinellidae, Coleoptera
Hipposiderobia *Hipposiderobia cloeotis* Fain *Hipposiderobia coelopos* Uchikawa	Myobiidae, Acari
Hirosia *Hirosia humilis* (Coquillett) *Hirosia iyoensis* (Shiraki) *Hirosia sapporoensis* (Shiraki)	Tabanidae, Diptera
Hirstia *Hirstia chelidonis* Hull *Hirstia domicola* Fain, Oshima & van Bronswijk	Pyroglyphidae, Acari
Hirstiella *Hirstiella trombidiiformis* (Berlese)	Pterygosomatidae, Acari
Hirstionyssus *Hirstionyssus apodemi* Zuevskiï *Hirstionyssus apoensis* Delfinado *Hirstionyssus blanchardi* (Trouessart) *Hirstionyssus confucianus* (Hirst) *Hirstionyssus criceti* (Sulzer) *Hirstionyssus cynomys* (Radford) *Hirstionyssus davydovae* Nikol'skiï *Hirstionyssus gansuensis* Ma & Piao	Laelapidae, Acari **see** *Hirstionyssus sunci*

Hirstionyssus huangheensis Ma & Piao
Hirstionyssus isabellinus (Oudemans)
Hirstionyssus meridianus Zemskaya
Hirstionyssus pratensis Gu & Yang
Hirstionyssus punctatus Gu & Yang
Hirstionyssus qinghaiensis Gu & Yang
Hirstionyssus sunci Wang
Hirstionyssus talpae Zemskaya
Hirstionyssus xinghaiensis Ma & Piao

Hirsutiella   Trombiculidae, Acari
Hirsutiella multisetosa (Willmann)   **see** *Hirsutiella zachvatkini*
Hirsutiella zachvatkini (Shluger)

Hister   Histeridae, Coleoptera
Hister abbreviatus Fabricius
Hister cadaverinus Hoffmann   **see** *Hister impressus*
Hister impressus Fabricius
Hister nomas Erichson
Hister quadrinotatus Scriba

Histiostoma   Histiostomatidae, Acari
Histiostoma dactyloscopicum Mahunka & Eraky
Histiostoma feroniarum (Dufour)   **see** *Anoetus feroniarum*
Histiostoma magdolnae Mahunka & Eraky
Histiostoma woolleyi Mahunka & Eraky

Hoffmanniella   Trombiculidae, Acari
Hoffmanniella transylvanica Goff

Hoffmannina   Trombiculidae, Acari
Hoffmannina danieli (Kolebinova)
Hoffmannina peruensis Goff
Hoffmannina raissae Gushcha & Kharadov
Hoffmannina theodori Gushcha
Hoffmannina tokobajevi Gushcha & Kharadov
Hoffmannina tshatkalica Gushcha & Kharadov

Hohorstiella   Menoponidae, Phthiraptera
Hohorstiella gigantea gigantea (Denny)
Hohorstiella gigantea lata (Piaget)   **see** *Hohorstiella lata*
Hohorstiella lata (Piaget)
Hohorstiella radovskyi Price & Emerson
Hohorstiella tenorioae Price & Emerson

Holakartikos   **see** *Bovicola*
Holakartikos crassipes (Rudow)   **see** *Bovicola crassipes*

Holobomolochus   Bomolochidae, Copepoda
Holobomolochus confusus (Stock)

Hololena   Agelenidae, Araneae
Hololena curta (McCook)

Holomenopon   Ancistromidae, Phthiraptera
Holomenopon maxbeieri Eichler

Holoparasitus   Parasitidae, Acari
Holoparasitus tirolensis (Sellnick)

Holophryxus   Dajidae, Isopoda
Holophryxus acanthephyrae Stephensen
Holophryxus polyandrus (Schultz)
Holophryxus quadratohumerale (Schultz)

Holotrichia	Scarabaeidae, Coleoptera
Holotrichia morosa Waterhouse	
Holotrichia oblita (Falderman)	
Holotrichia parallela (Motschulsky)	
Holubicula	Trombiculidae, Acari
Holubicula toroensis Daniel & Vercammen-Grandjean	
Hoplopleura	Hoplopleuridae, Phthiraptera
Hoplopleura acanthopus (Burmeister)	
Hoplopleura affinis (Burmeister)	
Hoplopleura aitkeni Johnson	
Hoplopleura andina Castro	
Hoplopleura cooki Kim	
Hoplopleura delticola Castro	
Hoplopleura edentula Fahrenholz	
Hoplopleura erratica (Osborn)	
Hoplopleura hirsuta Ferris	
Hoplopleura imparata Linardi, Teixeira & Botelho	
Hoplopleura ingens Castro	
Hoplopleura karachiensis Khanum	
Hoplopleura khandala Mishra	
Hoplopleura kondana Mishra	
Hoplopleura mendozana Castro	
Hoplopleura minasensis Linardi, Teixeira & Botelho	
Hoplopleura ochotonae Ferris	
Hoplopleura oenomydis Ferris	
Hoplopleura oxymycteri Ferris	
Hoplopleura ramgarh Mishra, Bhat & Kulkarni	
Hoplopleura sahyadri Mishra	
Hoplopleura sciuricola Ferris	
Hoplopleura similis Kim	
Hoplopleura sinhgarh Mishra, Bhat & Kulkarni	
Hoplopleura travassosi Werneck	
Hoplopsyllus	Pulicidae, Siphonaptera
Hoplopsyllus affinis (Baker)	**see** Euhoplopsyllus glacialis affinis
Hortensia	Cicadellidae, Hemiptera
Hortensia similis (Walker)	
Huckettomyia	Muscidae, Diptera
Huckettomyia watanabei Pont & Shinonaga	
Hughesiella	Pyroglyphidae, Acari
Hughesiella africana (Hughes)	
Hunterellus	**see** Ixodiphagus
Hunterellus hookeri Howard	**see** Ixodiphagus hookeri
Hunterellus sagarensis Geevarghese	**see** Ixodiphagus sagarensis
Hyalella	Hyalellidae, Amphipoda
Hyalella azteca (Saussure)	
Hyalomma	Ixodidae, Acari
Hyalomma aegyptium (Linnaeus)	
Hyalomma anatolicum anatolicum Koch	
Hyalomma anatolicum excavatum Koch	
Hyalomma arabica Pegram, Hoogstraal & Wassef	
Hyalomma asiaticum asiaticum Schulze & Schlottke	
Hyalomma asiaticum caucasicum Pomerantsev	
Hyalomma asiaticum kozlovi Olenev	**see** Hyalomma kozlovi
Hyalomma brevipunctata Sharif	
Hyalomma detritum Schulze	

Hyalomma dromedarii Koch
Hyalomma dromedarii asiaticum Schulze & Schlottke — **see** Hyalomma asiaticum
Hyalomma excavatum Koch — **see** Hyalomma anatolicum excavatum
Hyalomma hussaini Sharif
Hyalomma impeltatum Schulze & Schlottke
Hyalomma impressum Koch
Hyalomma kozlovi Olenev
Hyalomma lusitanicum Koch
Hyalomma marginatum isaaci Sharif
Hyalomma marginatum marginatum Koch
Hyalomma marginatum rufipes Koch
Hyalomma marginatum turanicum Pomerantsev
Hyalomma nitidum Schulze
Hyalomma plumbeum auct. — **see** Hyalomma marginatum
Hyalomma plumbeum asiaticum Schulze & Schlottke — **see** Hyalomma asiaticum
Hyalomma punt Hoogstraal, Kaiser & Pedersen
Hyalomma rhipicephaloides Neumann
Hyalomma rufipes Koch — **see** Hyalomma marginatum rufipes
Hyalomma schulzei Olenev
Hyalomma scupense Schulze — **see** Hyalomma detritum
Hyalomma truncatum Koch

Hybomitra — Tabanidae, Diptera
Hybomitra aequetincta (Becker)
Hybomitra agora Teskey, Shemanchuk & Weintraub
Hybomitra arpadi (Szilady)
Hybomitra bimaculata (Macquart)
Hybomitra borealis (Fabricius)
Hybomitra caucasica (Enderlein)
Hybomitra ciureai (Séguy)
Hybomitra distinguenda (Verrall)
Hybomitra epistates (Osten Sacken)
Hybomitra erberi (Brauer)
Hybomitra expollicata (Pandelle)
Hybomitra frontalis (Walker)
Hybomitra hearlei (Philip)
Hybomitra hinei hinei (Johnson)
Hybomitra hinei wrighti (Whitney)
Hybomitra hirticeps (Loew) — **see** Hybomitra lurida
Hybomitra illota (Osten Sacken)
Hybomitra lapponica (Wahlberg) — **see** Hybomitra borealis
Hybomitra lasiophthalma (Macquart)
Hybomitra laticornis (Hine) — **see** Tabanus laticornis
Hybomitra lundbecki Lyneborg
Hybomitra lurida (Fallén)
Hybomitra micans (Meigen)
Hybomitra montana (Meigen)
Hybomitra muehlfeldi (Brauer)
Hybomitra nitidifrons Szilady
Hybomitra nuda (McDunnough)
Hybomitra olsoi Takahasi
Hybomitra opaca (Coquillett)
Hybomitra peculiaris (Szilady)
Hybomitra schineri Lyneborg — **see** Hybomitra ciureai
Hybomitra sexfasciata (Hine)
Hybomitra sodalis (Williston)
Hybomitra tarandina (Linnaeus)
Hybomitra tropica (Linnaeus)
Hybomitra zonalis (Kirby)

Hybopygia — Sarcophagidae, Diptera
Hybopygia varia (Walker)

Hydaticus — Dytiscidae, Coleoptera

Hydrochara *Hydrochara affinis* Sharp	Hydrophilidae, Coleoptera
Hydrodroma *Hydrodroma despiciens despiciens* (Müller) *Hydrodroma despiciens pilosa* Besseling	Hydrodromidae, Acari
Hydromya	Sciomyzidae, Diptera
Hydrophilus *Hydrophilus affinis* (Sharp) *Hydrophilus triangularis* Say	Hydrophilidae, Coleoptera **see** *Hydrochara affinis*
Hydroporus *Hydroporus palustris* (Linnaeus) *Hydroporus undulatus* Say	Dytiscidae, Coleoptera
Hydrotaea *Hydrotaea aenescens* (Wiedemann) *Hydrotaea albipuncta* (Zetterstedt) *Hydrotaea armipes* (Fallén) *Hydrotaea capensis* Wiedemann *Hydrotaea dentipes* (Fabricius) *Hydrotaea hirtitibia* Stein *Hydrotaea ignava* (Harris) *Hydrotaea irritans* (Fallén) *Hydrotaea meridionalis* Portschinsky *Hydrotaea meteorica* (Linnaeus) *Hydrotaea pandellei* Stein *Hydrotaea pellucens* Portschinsky *Hydrotaea solitaria* Albuquerque	Muscidae, Diptera
Hylemya *Hylemya cinerella* (Fallén) *Hylemya strenua* Robineau-Desvoidy *Hylemya strigosa* (Fabricius)	Anthomyiidae, Diptera **see** *Paregle cinerella* **see** *Hylemya strenua*
Hylesia *Hylesia iola* Dyar *Hylesia lineata* Druce *Hylesia metabus* (Cramer) *Hylesia urticans* Floch & Abonnenc	Saturniidae, Lepidoptera
Hyperlaelaps *Hyperlaelaps arvalis* (Zakhvatkin) *Hyperlaelaps microti* (Ewing)	Laelapidae, Acari
Hyphantria *Hyphantria cunea* (Drury)	Arctiidae, Lepidoptera
Hypoaspis *Hypoaspis heselhausi* Oudemans *Hypoaspis lubricus* Voigts & Oudemans *Hypoaspis miles* (Berlese)	Laelapidae, Acari
Hypochrosis *Hypochrosis hyadaria* Guenée	Geometridae, Lepidoptera
Hypodectes *Hypodectes propus bulbuci* Fain *Hypodectes propus propus* (Nitzsch)	Hypoderatidae, Acari

Hypoderma	Oestridae, Diptera
Hypoderma actaeon Brauer	
Hypoderma bovis (Linnaeus)	
Hypoderma capreola Rubtsov	
Hypoderma diana Brauer	
Hypoderma lineatum lineatum (Villers)	
Hypoderma lineatum sinense Pleske	**see** *Hypoderma sinense*
Hypoderma sinense Pleske	
Hypoderma tarandi (Linnaeus)	
Hypogastrura	Hypogastruridae, Collembola
Hypogastrura scotti Yosii	
Hypoponera	Formicidae, Hymenoptera
Hypoponera eduardi (Forel)	
Hystrichopsylla	Hystrichopsyllidae, Siphonaptera
Hystrichopsylla orientalis Smit	
Hystrichopsylla talpae (Curtis)	
Hystrichopsylla weida qinlingensis Zhang, Wu & Liu	
Hystrichopsylla weida weida Jameson & Hsieh	
Hystricocnema	Sarcophagidae, Diptera
Hystricocnema plinthopyga (Wiedemann)	

I

Ichoronyssus	Macronyssidae, Acari
Ichoronyssus scutatus (Kolenati)	
Icosta	Hippoboscidae, Diptera
Icosta americana (Leach)	
Icosta ardeae (Macquart)	
Idiella	Calliphoridae, Diptera
Idiella tripartita (Bigot)	
Ilione	Sciomyzidae, Diptera
Ilione albiseta (Scopoli)	**see** *Knutsonia albiseta*
Ilyocoris	Naucoridae, Hemiptera
Ilyocoris cimicoides (Linnaeus)	
Indocentor	**see** *Dermacentor*
Indocentor ater Schulze	**see** *Dermacentor steini*
Iomachus	Ischnuridae, Scorpiones
Iomachus surgani Bastawade	
Irianopus	Acaridae, Acari
Irianopus brevis Fain	
Iridomyrmex	Formicidae, Hymenoptera
Iridomyrmex humilis (Mayr)	
Ischnopsyllus	Ischnopsyllidae, Siphonaptera
Ischnopsyllus delectabilis Smit	
Ischnopsyllus hexactenus (Kolenati)	
Ischnopsyllus intermedius (Rothschild)	
Ischnopsyllus jinciensis Xiao	
Ischnopsyllus magnabulga Xie, Yang & Li	
Ischnopsyllus octactenus (Kolenati)	
Ischnopsyllus petropolitanus (Wagner)	
Ischnopsyllus plumatus Ioff	
Ischnopsyllus qradrasetus Xie, Yang & Li	

Ischnopsyllus quintuesetus Xie, Yang & Li
Ischnopsyllus simplex Rothschild
Ischnopsyllus ussuriensis Medvedev
Ischnopsyllus variabilis (Wagner)

Ischnura   Coenagriidae, Odonata
Ischnura cervula Selys
Ischnura elegans (van der Linden)

Isometrus   Buthidae, Scorpiones
Isometrus maculatus (DeGeer)

Isomolgus   Lichomolgidae, Copepoda
Isomolgus desmotes Dojiri

Isomyia   Calliphoridae, Diptera

Isophryxus   **see** *Holophryxus*
Isophryxus polyandrus Schultz   **see** *Holophryxus polyandrus*
Isophryxus quadratohumerale Schultz   **see** *Holophryxus quadratohumerale*

Isshikia   Tabanidae, Diptera
Isshikia wenchuanensis Wang

Ixeuticus   Amaurobiidae, Araneae
Ixeuticus martius (Simon)
Ixeuticus robustus (L. Koch)

Ixobioides   Ixodorrhynchidae, Acari
Ixobioides brachispinosus Lizaso
Ixobioides butantanensis Fonseca
Ixobioides fonsecae (Fain)

Ixodes   Ixodidae, Acari
Ixodes acuminatus Neumann
Ixodes affinis Neumann
Ixodes amarali Fonseca
Ixodes anatis Chilton
Ixodes angustus Neumann
Ixodes apronophorus Schulze
Ixodes arboricola Schulze & Schlottke
Ixodes asanumai Kitaoka
Ixodes aulacodi Arthur
Ixodes baergi Cooley & Kohls
Ixodes brunneus Koch
Ixodes caledonicus Nuttall
Ixodes canisuga Johnston
Ixodes cavipalpus Nuttall & Warburton
Ixodes ceylonensis Kohls
Ixodes confusus Roberts
Ixodes cookei Packard
Ixodes cordifer Neumann
Ixodes cornuatus Roberts
Ixodes crenulatus Koch
Ixodes dammini Spielman *et al.*
Ixodes daveyi Nuttall
Ixodes dendrolagi Wilson
Ixodes dentatus Marx
Ixodes downsi Kohls
Ixodes drakensbergensis Clifford *et al.*
Ixodes elongatus Bedford
Ixodes eudyptidis Maskell
Ixodes euplecti Arthur

Ixodes festai Rondelli
Ixodes frontalis (Panzer)
Ixodes ghilarovi Filippova & Panova
Ixodes granulatus Supino
Ixodes hexagonus Leach
Ixodes holocyclus Neumann
Ixodes hyatti Clifford, Hoogstraal & Kohls
Ixodes kaschmiricus Serdyukova
Ixodes kingi Bishopp
Ixodes laguri Olenev
Ixodes lividus Koch
Ixodes marmotae Cooley & Kohls
Ixodes marxi Banks
Ixodes matopi Spickett, Keirans, Norval & Clifford
Ixodes melicola Schulze & Schlottke
Ixodes moreli Arthur
Ixodes moscharius Teng
Ixodes moschiferi Nemenz
Ixodes muris Bishopp & Smith
Ixodes myospalacis Teng
Ixodes neitzi Clifford, Walker & Keirans
Ixodes neotomae Cooley
Ixodes nipponensis Kitaoka & Saito
Ixodes ovatus Neumann
Ixodes pacificus Cooley & Kohls
Ixodes pararicinus Keirans & Clifford
Ixodes passericola Schulze
Ixodes pavlovskyi Pomerantsev
Ixodes persulcatus Schulze
Ixodes petauristae Warburton
Ixodes pilosus Koch
Ixodes pomerantzevi Serdyukova
Ixodes pseudorasus Arthur & Burrow
Ixodes putus (Pickard–Cambridge)   **see** *Ixodes uriae*
Ixodes redikorzevi Olenev
Ixodes ricinus (Linnaeus)
Ixodes rothschildi Nuttall & Warburton
Ixodes rubicundus Neumann
Ixodes scapularis Say
Ixodes schillingsi Neumann
Ixodes siamensis Kitaoka & Suzuki
Ixodes simplex Neumann
Ixodes sinensis Teng
Ixodes spinicoxalis Neumann
Ixodes tanuki Saito
Ixodes tasmani Neumann
Ixodes texanus Banks
Ixodes thomasae Arthur & Burrow
Ixodes trianguliceps Birulya
Ixodes turdus Nakatsuji
Ixodes ugandanus Neumann
Ixodes unicavatus Neumann
Ixodes uriae White
Ixodes ventalloi Gil Collado
Ixodes vespertilionis Koch
Ixodes walkerae Clifford, Kohls & Hoogstraal
Ixodes woodi Bishopp

Ixodiphagus   Encyrtidae, Hymenoptera
Ixodiphagus hookeri (Howard)
Ixodiphagus sagarensis (Geevarghese)
Ixodiphagus texanus Howard

J

Johnstonimyia
Johnstonimyia imitatrix Lopes

Sarcophagidae, Diptera
see Sarcophaga imitatrix

K

Kelerinirmus
Kelerinirmus fuscus (Denny)

Degeeriellidae, Phthiraptera

Kheper
Kheper devotus (Redtenbacher)
Kheper lamarcki (Macleay)

Scarabaeidae, Coleoptera

Kiefferulus
Kiefferulus barbatitarsis (Kieffer)

Chironomidae, Diptera

Kiricephalus
Kiricephalus pattoni (Stephens)

Linguatulidae, Pentastomida

Kleidotoma
Kleidotoma psiloides Westwood

Eucoilidae, Hymenoptera

Knemidokoptes
Knemidokoptes gallinae (Railliet)
Knemidokoptes intermedius Fain & Macfarlane
Knemidokoptes mutans (Robin & Lanquetin)
Knemidokoptes pilae Lavoipierre & Griffiths

Knemidokoptidae, Acari
see Neocnemidocoptes gallinae

Knutsonia
Knutsonia albiseta (Scopoli)
Knutsonia trifaria (Loew)

Sciomyzidae, Diptera

Koeniginirmus
Koeniginirmus sellatus Burmeister

Philopteridae, Phthiraptera
see Cummingsiella sellatus

Kogotus
Kogotus nonus (Needham & Claassen)

Perlidae, Plecoptera

Koptortosoma

see Xylocopa

Kramerea
Kramerea schuetzei (Kramer)

Sarcophagidae, Diptera

Kroeyerina
Kroeyerina benzorum Deets
Kroeyerina cortezensis Deets
Kroeyerina deborahae Deets
Kroeyerina mobulae Deets

Kroyeriidae, Copepoda

L

Labia
Labia cauvicauda Motschulsky
Labia minor (Linnaeus)

Labiidae, Dermaptera

Labidophorus
Labidophorus nearcticus Fain & Whitaker

Glycyphagidae, Acari

Lacconectus
Lacconectus punctipennis Zimmermann

Dytiscidae, Coleoptera

Laccophilus
Laccophilus maculosus (Germar)

Dytiscidae, Coleoptera

Laccotrephes *Laccotrephes griseus* (Guérin-Méneville) *Laccotrephes ruber* (Linnaeus)	Nepidae, Hemiptera
Lachnosterna	**see** *Holotrichia* or *Phyllophaga*
Lachnus *Lachnus roboris* (Linnaeus)	Aphididae, Hemiptera
Laelaps *Laelaps agilis* Koch *Laelaps algericus* Hirst *Laelaps alongensis* Grokhovskaya & Nguen Huan Hoe *Laelaps benoiti* Taufflieb *Laelaps clethrionomydis* Lange *Laelaps echidnina* Berlese *Laelaps hilaris* Koch *Laelaps jettmari* Vitzthum *Laelaps kochi* Oudemans *Laelaps lativentralis* Fonseca *Laelaps liberiensis* Hirst *Laelaps manguinhosi* Fonseca *Laelaps muris* (Ljungh) *Laelaps nigeriensis* Keegan *Laelaps nuttalli* Hirst *Laelaps paulistanensis* Fonseca *Laelaps pavlovskyi* Zakhvatkin *Laelaps roubaudi* Taufflieb *Laelaps traubi* Domrow	Laelapidae, Acari **see** *Hyperlaelaps microti*
Laemobothrion *Laemobothrion iberum* Pérez–Jiménez *et al.* *Laemobothrion maximum* Scopoli	Laemobothriidae, Phthiraptera
Laetulonthus	Staphylinidae, Coleoptera
Lagaropsylla *Lagaropsylla anciauxi* Smit *Lagaropsylla hoogstraali* Smit *Lagaropsylla mytila* Hůrka	Ischnopsyllidae, Siphonaptera
Lagochondria *Lagochondria nana* Ho & Dojiri	Chondracanthidae, Copepoda
Lagopoecus *Lagopoecus colchicus* Emerson	Degeeriellidae, Phthiraptera
Laminosioptes *Laminosioptes cysticola* (Vizioli)	Laminosioptidae, Acari
Lampona *Lampona cylindrata* (L. Koch)	Gnaphosidae, Araneae
Lamproglena *Lamproglena monodi* Capart	Lernaeidae, Copepoda
Lardoglyphus *Lardoglyphus konoi* (Sasa & Asanuma)	Lardoglyphidae, Acari
Laronyssus *Laronyssus martini* (Trouessart)	Pterolichidae, Acari
Lasioderma *Lasioderma serricorne* (Fabricius)	Anobiidae, Coleoptera

Lasiodora *Lasiodora erythrocythara* Mello–Leitão	Theraphosidae, Araneae
Lasioglossum *Lasioglossum zephyrum* (Smith)	Apidae, Hymenoptera
Lasiohelea *Lasiohelea stimulans* (de Meijere)	**see** *Forcipomyia* **see** *Forcipomyia stimulans*
Lasius *Lasius alienus* (Förster) *Lasius fuliginosus* (Latreille) *Lasius niger* (Linnaeus)	Formicidae, Hymenoptera
Latrodectus *Latrodectus atritus* Urquhart *Latrodectus curacaviensis* (Müller) *Latrodectus dahli* Levi *Latrodectus geometricus* Koch *Latrodectus hasseltii* Thorell *Latrodectus hesperus* Chamberlin & Ivie *Latrodectus indistinctus* Pickard–Cambridge *Latrodectus katipo* Powell *Latrodectus mactans* (Fabricius) *Latrodectus mactans tredecimguttatus* (Rossi) *Latrodectus menavodi* Vinson *Latrodectus pallidus pallidus* Pickard–Cambridge *Latrodectus pallidus pavlovskii* Kharitonov *Latrodectus revivensis* Shulov *Latrodectus tredecimguttatus* (Rossi) *Latrodectus variolus* Walckenaer	Theridiidae, Araneae **see** *Latrodectus tredecimguttatus*
Lawrenceocarpus *Lawrenceocarpus chilonycteris* Fain	Chirodiscidae, Acari
Leeuwenhoekia *Leeuwenhoekia australiensis* Hirst	Leeuwenhoekiidae, Acari **see** *Odontacarus australiensis*
Leiodinychus *Leiodinychus krameri* (G. & R. Canestrini)	Uropodidae, Acari
Leiperia *Leiperia gracile* (Diesing)	Porocephalidae, Pentastomida
Leistotrophus *Leistotrophus versicolor* (Gravenhorst)	Staphylinidae, Coleoptera
Leiurus *Leiurus quinquestriatus hebraeus* (Birulya) *Leiurus quinquestriatus quinquestriatus* (Hemprich & Ehrenberg)	Buthidae, Scorpiones
Lepeophtheirus *Lepeophtheirus salmonis* (Krøyer)	Caligidae, Copepoda
Lepidoglyphus *Lepidoglyphus destructor* (Schrank)	Glycyphagidae, Acari
Lepikentron *Lepikentron ovis* (Schrank)	Trichodectidae, Phthiraptera **see** *Bovicola ovis*
Lepinotus *Lepinotus patruelis* Pearman	Trogiidae, Psocoptera

Lepisma *Lepisma saccharina* Linnaeus	Lepismatidae, Thysanura
Lepismodes	**see** *Thermobia*
Leporacarus *Leporacarus gibbus* (Pagenstecher) *Leporacarus sylvilagi* Fain, Whitaker & Lukoschus	Listrophoridae, Acari
Leptacinus *Leptacinus socius* (Fauvel)	Staphylinidae, Coleoptera
Leptagrion *Leptagrion siqueirai* Santos	Coenagriidae, Odonata
Leptocera *Leptocera acutangula* (Zetterstedt) *Leptocera caenosa* (Rondani) *Leptocera digna* Roháček *Leptocera ferruginata* (Stenhammer) *Leptocera fontinalis* (Fallén) *Leptocera fuscipennis* (Haliday) *Leptocera hirtula* (Rondani) *Leptocera vagans* (Haliday)	Sphaeroceridae, Diptera **see** *Coproica acutangula* **see** *Coproica ferruginata* **see** *Coproica hirtula* **see** *Coproica vagans*
Leptoconops *Leptoconops bahreinensis* Clastrier & Boorman *Leptoconops bequaerti* (Kieffer) *Leptoconops demeilloni* Clastrier & Nevill *Leptoconops foulki* Clastrier & Wirth *Leptoconops hamariensis* (Herzi & Sabatini) *Leptoconops kerteszi* Kieffer *Leptoconops knowltoni* Clastrier & Wirth *Leptoconops laurae* (Weiss) *Leptoconops mediterraneus* (Kieffer) *Leptoconops mellori* Clastrier & Boorman *Leptoconops nipponensis nipponensis* Tokunaga *Leptoconops nipponensis oshimaensis* Takaoka & Hayashi *Leptoconops spinosifrons* (Carter) *Leptoconops torrens* (Townsend)	Ceratopogonidae, Diptera **see** *Leptoconops laurae*
Leptocyclopodia *Leptocyclopodia ferrarii* (Rondani)	Nycteribiidae, Diptera
Leptolichus	Eustathiidae, Acari
Leptopsylla *Leptopsylla pectiniceps pectiniceps* (Wagner) *Leptopsylla pectiniceps ventrisinulata* Chen, Zhang & Liu *Leptopsylla segnis* (Schönherr) *Leptopsylla silvatica* (Meinert)	Leptopsyllidae, Siphonaptera **see** *Peromyscopsylla silvatica*
Leptothorax *Leptothorax curvispinosus* Mayr	Formicidae, Hymenoptera
Leptotrombidium *Leptotrombidium akamushi* (Brumpt) *Leptotrombidium alpinum* Yu & Yang *Leptotrombidium andrei* Tanskul & Gingrich *Leptotrombidium arenicola* Traub *Leptotrombidium bishanense* Yu, Hu & Fang *Leptotrombidium deliense deliense* (Walch) *Leptotrombidium deliense microsetosa* Zhao, Tang & Mo *Leptotrombidium dichotogalium* Xiang & Wen	Trombiculidae, Acari

Leptotrombidium eothenomydis Yu & Yang
Leptotrombidium europaea (Daniel & Brelih)
Leptotrombidium fletcheri (Womersley & Heaslip)
Leptotrombidium fujianensis Liao & Wang
Leptotrombidium gaohuensis Wei, Tong & Shi
Leptotrombidium guzhangensis Wang, Li & Tian
Leptotrombidium harrisoni Tanskul & Gingrich
Leptotrombidium heishuiense Zhou, Chen & Wang
Leptotrombidium himizui Sasa *et al.*
Leptotrombidium hsui Yu, Yang & Gong
Leptotrombidium intermedium (Nagayo *et al.*)
Leptotrombidium kalyani Kulkarni
Leptotrombidium kaohuense (Yang *et al.*)
Leptotrombidium kitasatoi (Fukuzumi & Obata)
Leptotrombidium laojunshanense Yu, Yang & Gong
Leptotrombidium longisetum (Womersley)
Leptotrombidium longisetum Yu, Hu & Fang   **see** *Leptotrombidium bishanense*
Leptotrombidium miyazakii (Sasa *et al.*)
Leptotrombidium neotebraci Xiang & Wen
Leptotrombidium ochotonae Wang, Zhai & Chen
Leptotrombidium orestes Xiang & Wen
Leptotrombidium owuense (Sasa & Kumada)
Leptotrombidium pallidum (Nagayo *et al.*)
Leptotrombidium palpale (Nagayo *et al.*)
Leptotrombidium pilosum Traub & Lakshana
Leptotrombidium qiui Yu, Yang & Gong
Leptotrombidium quanzhouensis (Liao, Lin & Wang)
Leptotrombidium rajaniae Kulkarni
Leptotrombidium rubellum Wang & Liao
Leptotrombidium rusticum Yu, Yang & Gong
Leptotrombidium scutellare (Nagayo *et al.*)
Leptotrombidium sokolovi Kudryashova
Leptotrombidium spicanisetum Yu, Yang & Gong
Leptotrombidium spinifoliatum Wang, Li & Tian
Leptotrombidium taishanicum Meng, Xue & Wen
Leptotrombidium wangi Yu, Yang & Gong
Leptotrombidium yongshengense Yu & Yang

Leptus   Erythraeidae, Acari
Leptus irritans Riley   **see** *Eutrombicula alfreddugesi*

Lernaea   Lernaeidae, Copepoda
Lernaea catostomi Krøyer
Lernaea cyprinacea Linnaeus
Lernaea elegans Leigh-Sharpe
Lernaea lophiara Harding

Lernaeenicus   Pennellidae, Copepoda
Lernaeenicus encrasicoli (Turton)
Lernaeenicus longiventris Wilson
Lernaeenicus polyceraus Wilson
Lernaeenicus radiatus Le Sueur
Lernaeenicus sprattae (Sowerby)

Lernaeocera   Pennellidae, Copepoda
Lernaeocera branchialis (Linnaeus)

Lernaeopoda   Lernaeopodidae, Copepoda

Lernanthropinus   Lernanthropidae, Copepoda
Lernanthropinus nematistii Deets & Benz

Lernanthropus   Lernanthropidae, Copepoda
Lernanthropus atrox Heller

Lethocerus *Lethocerus deyrollei* (Vuillefroy) *Lethocerus indicus* (Lepeletier & Serville)	Belostomatidae, Hemiptera
Leucophaea *Leucophaea maderae* (Fabricius)	Oxyhaloidae, Dictyoptera **see** *Rhyparobia maderae*
Leucotabanus *Leucotabanus pauculus* Fairchild	Tabanidae, Diptera
Levitinia *Levitinia freidbergi* Beaucournu-Saguez & Braverman *Levitinia tacobi* Chubareva & Petrova	Simuliidae, Diptera
Liatonga	**see** *Liatongus*
Liatongus *Liatongus militaris* (Laporte) *Liatongus phanaeoides* (Westwood)	Scarabaeidae, Coleoptera
Libellula *Libellula luctuosa* Burmeister *Libellula lydia* Drury	Libellulidae, Odonata
Liberonautes *Liberonautes latidactylus latidactylus* (Deman) *Liberonautes latidactylus nanoides* Cumberlidge & Sachs *Liberonautes latidactylus paludicolis* Cumberlidge & Sachs *Liberonautes latidactylus rubigimanus* Cumberlidge & Sachs	Potamonautidae, Decapoda
Liebstadia *Liebstadia similis* (Michael)	Scheloribatidae, Acari
Limatus *Limatus durhamii* Theobald *Limatus flavisetosus* de Oliveira Castro	Culicidae, Diptera
Limnia *Limnia unguicornis* (Scopoli)	Sciomyzidae, Diptera
Limnophora *Limnophora patellifera* Villeneuve	Muscidae, Diptera
Limnophyes *Limnophyes hudsoni* Saether *Limnophyes minimus* (Meigen)	Chironomidae, Diptera
Linguatula *Linguatula arctica* Riley, Haugerud & Nilssen *Linguatula denticulata* (Rudolphi) *Linguatula dingophila* Johnson *Linguatula recurvata* (Diesing) *Linguatula serrata* Frölich	Linguatulidae, Pentastomida **see** *Linguatula serrata*
Linognathoides *Linognathoides cynomyis* Kim	Polyplacidae, Phthiraptera
Linognathus *Linognathus africanus* Kellogg & Paine *Linognathus armatus* Fiedler & Stampa *Linognathus euchore* Waterston *Linognathus fahrenholzi* Paine *Linognathus oviformis* Rudow *Linognathus ovillus* (Neumann)	Linognathidae, Phthiraptera

Linognathus pedalis (Osborn)
Linognathus setosus (von Olfers)
Linognathus stenopsis (Burmeister)
Linognathus vituli (Linnaeus)

Liohippelates   Chloropidae, Diptera
Liohippelates bishoppi (Sabrosky)
Liohippelates collusor (Townsend)
Liohippelates pallipes (Loew)
Liohippelates pusio (Loew)

Liopygia   **see** *Sarcophaga*
Liopygia ruficornis (Fabricius)   **see** *Sarcophaga ruficornis*

Lipeurus   Philopteridae, Phthiraptera
Lipeurus caponis (Linnaeus)
Lipeurus lawrensis Bedford   **see** *Numidilipeurus lawrensis*
Lipeurus maculosus Clay
Lipeurus parkeri Emerson & Price
Lipeurus tropicalis Peters   **see** *Numidilipeurus lawrensis tropicalis*

Liponyssoides   Dermanyssidae, Acari
Liponyssoides sanguineus (Hirst)

Liponyssus   **see** *Ornithonyssus*
Liponyssus sylviarum (Canestrini & Fanzago)   **see** *Ornithonyssus sylviarum*

Lipoptena   Hippoboscidae, Diptera
Lipoptena capreoli Rondani
Lipoptena cervi (Linnaeus)
Lipoptena depressa (Say)
Lipoptena mazamae Rondani

Liposcelis   Liposcelidae, Psocoptera
Liposcelis pubescens Broadhead

Lironeca   Cymothoidae, Isopoda
Lironeca atlantniroi Kononenko

Lispe   Muscidae, Diptera
Lispe leucospila (Wiedemann)
Lispe orientalis Wiedemann

Listrophoroides   Atopomelidae, Acari
Listrophoroides kinabaluensis Fain

Listrophorus   Listrophoridae, Acari
Listrophorus gibbus Pagenstecher   **see** *Leporacarus gibbus*
Listrophorus mexicanus Fain
Listrophorus mustelae Mégnin   **see** *Lynxacarus mustelae*

Lithobius   Lithobiidae, Chilopoda
Lithobius forficatus (Linnaeus)
Lithobius pilicornis Newport

Liuopsylla   Hystrichopsyllidae, Siphonaptera
Liuopsylla conica Zhang, Wu & Liu

Livoneca   Cymothoidae, Isopoda

Lomis   Lithodidae, Decapoda
Lomis hirta Lamark

Loomisia *Loomisia peruviensis* Goff, Whitaker & Barkley	Trombiculidae, Acari
Lophoura *Lophoura tetraphylla* Ho	Sphyriidae, Copepoda
Lophyrotoma *Lophyrotoma interrupta* (Klug)	Tenthredinidae, Hymenoptera
Lorillatum	Trombiculidae, Acari
Lotophila *Lotophila atra* (Meigen)	Sphaeroceridae, Diptera
Lowryacarus *Lowryacarus longipes* Fain	Acaridae, Acari
Loxanoetus *Loxanoetus lenae* Domrow & Ladds	Anoetidae, Acari

Loxosceles   Loxoscelidae, Araneae
Loxosceles laeta (Nicolet)
Loxosceles parrami Newlands
Loxosceles reclusa Gertsch & Mulaik
Loxosceles rufipes (Lucas)
Loxosceles spiniceps Lawrence   **see** *Loxosceles spinulosa*
Loxosceles spinulosa Purcell

Lucilia   Calliphoridae, Diptera
Lucilia adisoemartoi Kurahashi
Lucilia ampullacea Villeneuve
Lucilia bismarckensis Kurahashi
Lucilia bufonivora Moniez
Lucilia caesar (Linnaeus)
Lucilia coeruleiviridis Macquart
Lucilia cuprina (Wiedemann)
Lucilia eximia (Wiedemann)
Lucilia ibis Shannon
Lucilia illustris (Meigen)
Lucilia mexicana Macquart
Lucilia sericata (Meigen)
Lucilia silvarum (Meigen)

Lukoschuscoptes   Knemidokoptidae, Acari
Lukoschuscoptes asiaticus OConnor *et al.*

Lutzomyia   Psychodidae, Diptera
Lutzomyia abonnenci (Floch & Chassignet)
Lutzomyia amilcari Arredondo Cardona
Lutzomyia anduzei (Rozeboom)
Lutzomyia anthophora (Addis)
Lutzomyia antunesi (Coutinho)
Lutzomyia atroclavata (Knab)
Lutzomyia aulari Feliciangeli, Ordonez & Manzanilla
Lutzomyia ayacuchensis Caceres & Galati
Lutzomyia ayrozai (Barretto & Coutinho)
Lutzomyia baculus Martins, Falcão & Silva
Lutzomyia baityi (Damasceno, Causey & Arouck)
Lutzomyia barrettoi (Mangabeira Filho)
Lutzomyia begonae (Ortiz & Torres)
Lutzomyia beltrani (Vargas & Díaz Nájera)
Lutzomyia beniensis Pont & Desjeux
Lutzomyia bernalei (Osorno-Mesa *et al.*)
Lutzomyia bettinii Feliciangeli *et al.*

Lutzomyia brisolai Pont & Desjeux
Lutzomyia capixaba Santos Dias et al.
Lutzomyia carmelinoi Ryan, Fraiha, Lainson & Shaw
Lutzomyia carrerai (Pereira Barretto)
Lutzomyia cayennensis (Floch & Abonnenc)
Lutzomyia cerqueirai (Causey & Damasceno)
Lutzomyia chagasi (Costa Lima)
Lutzomyia christophei (Fairchild & Trapido)
Lutzomyia claustrei Abonnenc, Léger & Fauran
Lutzomyia columbiana (Ristorcelli & van Ty)
Lutzomyia complexa (Mangabeira)
Lutzomyia cortelezzii (Brèthes)
Lutzomyia cruciata (Coquillett)
Lutzomyia cruzi (Mangabeira)
Lutzomyia davisi (Root)
Lutzomyia diabolica (Hall)
Lutzomyia disneyi Williams
Lutzomyia erwindonaldoi (Ortiz)
Lutzomyia evandroi (Costa Lima & Antunes)
Lutzomyia ferroae Young & Morales
Lutzomyia fischeri (Pinto)
Lutzomyia flaviscutellata (Mangabeira)
Lutzomyia forattinii Galati, Rego, Nunes & Teruya
Lutzomyia furcata (Mangabeira)
Lutzomyia gantieri Pont & Desjeux
Lutzomyia gasparviannai Martins, Godoy & Silva
Lutzomyia geniculata (Mangabeira)
Lutzomyia goiana Martins, Falcâo & Silva
Lutzomyia gomezi (Nitzulescu)
Lutzomyia guatemalensis Porter & Young
Lutzomyia hartmanni (Fairchild & Hertig)
Lutzomyia hirsuta (Mangabeira)
Lutzomyia intermedia (Lutz & Neiva)
Lutzomyia killicki Feliciangeli et al.
Lutzomyia lainsoni (Fraiha & Ward)
Lutzomyia lanei (Barretto & Coutinho)
Lutzomyia larensis Arredondo
Lutzomyia leonidasdeanei (Fraiha et al.)
Lutzomyia lewisi Feliciangeli, Ordonez & Fernandez
Lutzomyia llanosmartinsi (Fraiha & Ward)
Lutzomyia longipalpis (Lutz & Neiva)
Lutzomyia longispina (Mangabeira)
Lutzomyia marajoensis (Damasceno & Causey)   **see** Lutzomyia walkeri
Lutzomyia martinezi Young & Morales
Lutzomyia micropyga (Mangabeira)
Lutzomyia migonei (França)
Lutzomyia nordestina (Mangabeira)   **see** Lutzomyia sordellii
Lutzomyia nuneztovari anglesi Le Pont & Desjeux
Lutzomyia nuneztovari nuneztovari (Ortiz)
Lutzomyia oligodonta Young, Pérez & Romero
Lutzomyia olmeca bicolor Fairchild & Theodor
Lutzomyia olmeca nociva Young & Arias
Lutzomyia olmeca olmeca (Vargas & Díaz Nájera)
Lutzomyia olmeca reducta Feliciangeli et al.
Lutzomyia orestes (Fairchild & Trapido)
Lutzomyia ovallesi (Ortiz)
Lutzomyia panamensis (Shannon)
Lutzomyia paraensis (Costa Lima)
Lutzomyia pascalei (Coutinho & Pereira Barretto)
Lutzomyia peresi (Mangabeira)
Lutzomyia pessoai (Coutinho & Pereira Barretto)
Lutzomyia petropolitana Martins & Silva
Lutzomyia pinottii (Damasceno & Arouck)
Lutzomyia preclara Young & Arias

Lutzomyia punctigeniculata (Floch & Abonnenc)
Lutzomyia pusilla Dias, Martins, Falcão & da Silva
Lutzomyia rangeliana (Ortiz)
Lutzomyia ratcliffei Arias, Ready & de Freitas
Lutzomyia renei (Martins, Falcão & da Silva)
Lutzomyia rorotaensis (Floch & Abonnenc)
Lutzomyia sanguinaria (Fairchild & Hertig)
Lutzomyia scorzai (Ortiz)
Lutzomyia shannoni (Dyar)
Lutzomyia shawi Fraiha, Ward & Ready
Lutzomyia sordellii (Shannon & del Ponte)
Lutzomyia spinicrassa Morales-Alarcon *et al.*
Lutzomyia squamiventris maripaensis (Floch & Abonnenc)
Lutzomyia squamiventris squamiventris (Lutz & Neiva)
Lutzomyia tortura Young & Rogers
Lutzomyia townsendi (Ortiz)
Lutzomyia trapidoi (Fairchild & Hertig)
Lutzomyia trichopyga (Floch & Abonnenc)
Lutzomyia trinidadensis (Newstead)
Lutzomyia triramula (Fairchild & Hertig)
Lutzomyia ubiquitalis (Mangabeira)
Lutzomyia umbratilis Ward & Fraiha
Lutzomyia verrucarum (Townsend)
Lutzomyia vespertilionis (Fairchild & Hertig)
Lutzomyia vexator (Coquillett)
Lutzomyia walkeri (Newstead)
Lutzomyia waltoni Arias, de Freitas & Barrett
Lutzomyia wellcomei (Fraiha, Shaw & Lainson)
Lutzomyia whitmani (Antunes & Coutinho)
Lutzomyia williamsi (Damasceno, Causey & Arouck)
Lutzomyia witoto Young & Morales
Lutzomyia ylephiletrix (Fairchild & Hertig)
Lutzomyia youngi Feliciangeli & Murillo
Lutzomyia yucumensis (Le Pont *et al.*)
Lutzomyia yuilli pajoti Abonnenc, Léger & Fauran
Lutzomyia yuilli yuilli Young & Porter
Lutzomyia zeledoni Young & Murillo

Lychas	Buthidae, Scorpiones
Lychas laevifrons Pocock	
Lycosa	Lycosidae, Araneae
Lycosa erythrognata Lucas	
Lycosa godeffroyi L. Koch	
Lycosa indagastrix Walckenaer	
Lycosa raptoria Walckenaer	**see** *Scaptocosa raptoria*
Lycosa singoriensis (Laxmann)	
Lycosa vultuosa Koch	
Lyctocoris	Anthocoridae, Hemiptera
Lyctocoris campestris Fabricius	
Lynxacarus	Listrophoridae, Acari
Lynxacarus mustelae (Mégnin)	
Lynxacarus radovskyi Tenorio	**see** *Felistrophorus radovskyi*
Lyperosia	Muscidae, Diptera
Lyperosia irritans (Linnaeus)	**see** *Haematobia irritans*
Lyperosia minuta Bezzi	**see** *Haematobia minuta*
Lyperosia titillans Bezzi	**see** *Haematobia titillans*
Lyriothemis	Libellulidae, Odonata
Lyriothemis cleis Brauer	

Lytta	Meloidae, Coleoptera
Lytta neivai Denier	

M

Mabra	Pyralidae, Lepidoptera
Mabra elephantophila Bänziger	
Mabra haematophaga Bänziger	
Mabra lacriphaga Bänziger	

Macrocheles	Macrochelidae, Acari
Macrocheles adenostictus Krantz & Whitaker	
Macrocheles aestivus Halliday	
Macrocheles aethiopicus Berlese	
Macrocheles areolatus Krantz & Whitaker	
Macrocheles dolichosternus Krantz & Whitaker	
Macrocheles elimatus Berlese	
Macrocheles glaber (Müller)	
Macrocheles grossipes Berlese	
Macrocheles hirsutissima Berlese	
Macrocheles hirsutissima Evans & Hyatt	**see** *Macrocheles laciniatus*
Macrocheles inornatus Evans & Hyatt	**see** *Macrocheles elimatus*
Macrocheles kraepelini (Berlese)	
Macrocheles laciniatus Krantz	
Macrocheles longipes (Berlese)	
Macrocheles merdarius (Berlese)	
Macrocheles montanus (Willmann)	
Macrocheles muscaedomesticae (Scopoli)	
Macrocheles nataliae Bregetova & Koroleva	
Macrocheles nemontanus Ryke & Meyer	**see** *Macrocheles longipes*
Macrocheles peregrinus Krantz	
Macrocheles perglaber Filipponi & Pegazzano	
Macrocheles rotundiscutis Bregetova & Koroleva	
Macrocheles schaeferi Walter	
Macrocheles spinipes Berlese	
Macrocheles subbadius (Berlese)	

Macromia	Corduliidae, Odonata
Macromia magnifica Selys	

Macronema	Hydropsychidae, Trichoptera

Macronyssoides	**see** *Ornithonyssus*
Macronyssoides kochi (Fonseça)	**see** *Ornithonyssus kochi*

Macronyssus	Macronyssidae, Acari
Macronyssus flavus (Kolenati)	
Macronyssus laifengensis Wang & Shi	
Macronyssus miraspinosus Gu & Wang	
Macronyssus tashanensis Li & Teng	
Macronyssus zhijinensis Gu & Wang	

Macrostylophora	Ceratophyllidae, Siphonaptera
Macrostylophora angustihamulus Li, Zhang & Zeng	
Macrostylophora furcata Shi, Liu & Wu	

Maculinirmus	Philopteridae, Phthiraptera
Maculinirmus granatensis Soler Cruz et al.	

Magicicada	Cicadidae, Hemiptera
Magicicada septendecim (Linnaeus)	

Maladera	Scarabaeidae, Coleoptera
Maladera orientalis (Motschulsky)	

Malaraeus	Ceratophyllidae, Siphonaptera
Malaraeus euphorbi (Rothschild)	**see** *Amaradix euphorbi*
Malaraeus penicilliger (Grube)	**see** *Amalaraeus penicilliger*
Malaraeus telchinus (Rothschild)	

Malaya Culicidae, Diptera
Malaya genurostris Leicester

Malayoglyphus Pyroglyphidae, Acari
Malayoglyphus intermedius Fain, Cunnington & Spieksma

Mallos Dictynidae, Araneae
Mallos gregalis (Simon)

Mangora Araneidae, Araneae
Mangora pia Chamberlin & Ivie

Mansonia Culicidae, Diptera
Mansonia africana (Theobald)
Mansonia annulata Leicester
Mansonia annulifera (Theobald)
Mansonia bonneae Edwards
Mansonia dives (Schiner)
Mansonia dyari Belkin, Heinemann & Page
Mansonia fuscopennata (Theobald) **see** *Coquillettidia fuscopennata*
Mansonia indiana Edwards
Mansonia indubitans Dyar & Shannon
Mansonia nigricans (Coquillett) **see** *Coquillettidia nigricans*
Mansonia perturbans (Walker) **see** *Coquillettidia perturbans*
Mansonia richiardii (Ficalbi) **see** *Coquillettidia richiardii*
Mansonia titillans (Walker)
Mansonia uniformis (Theobald)

Margaropus Ixodidae, Acari
Margaropus winthemi Karsch

Mastotermes Mastotermitidae, Isoptera
Mastotermes darwiniensis Froggatt

Mayacnephia Simuliidae, Diptera
Mayacnephia aguirrei (Dalmat)
Mayacnephia fortunensis Petersen
Mayacnephia salasi Ramírez–Pérez *et al.*

Mecaderochondria Chondracanthidae, Copepoda
Mecaderochondria pilgrimi Ho & Dojiri

Megabothris Ceratophyllidae, Siphonaptera
Megabothris calcarifer (Wagner)
Megabothris clantoni clantoni Hubbard
Megabothris clantoni princei Hubbard
Megabothris turbidus (Rothschild)
Megabothris walkeri (Rothschild)

Megachernes Chernetidae, Pseudoscorpiones
Megachernes grandis (Beier)

Megachile Apidae, Hymenoptera
Megachile rotundata (Fabricius)

Megacyclops Cyclopidae, Copepoda
Megacyclops viridis (Jurine)

Megalopyge *Megalopyge opercularis* (J. E. Smith)	Megalopygidae, Lepidoptera
Megametope *Megametope carinatus* (Baker)	Xanthidae, Decapoda
Megapodiella *Megapodiella parkeri* Price & Emerson	Philopteridae, Phthiraptera
Megarcys	Perlodidae, Plecoptera
Megascolia *Megascolia flavifrons* (Fabricius)	Scoliidae, Hymenoptera
Megaselia *Megaselia abdita* Schmitz *Megaselia aurea* (Aldrich) *Megaselia biarticulata* Disney *Megaselia kurahashii* Disney *Megaselia scalaris* (Loew)	Phoridae, Diptera
Megathopa *Megathopa villosa* Eschscholtz	Scarabaeidae, Coleoptera
Megistopoda	Streblidae, Diptera
Megninia *Megninia crinita* Gaud, Atyeo & Barré *Megninia cubitalis* (Mégnin) *Megninia dipeltata* Gaud, Atyeo & Barré *Megninia ginglymura* (Mégnin) *Megninia hologastra* Gaud *Megninia ortari* Gaud, Atyeo & Barré	Analgidae, Acari
Melanogryllus *Melanogryllus desertus* (Pallas)	Gryllidae, Orthopetra
Melichares *Melichares longisetosus* (Postner)	Ascidae, Acari
Melinda *Melinda itoi* Kano	Calliphoridae, Diptera
Melittobia *Melittobia australica* Girault	Eulophidae, Hymenoptera
Melophagus *Melophagus ovinus* (Linnaeus)	Hippoboscidae, Diptera
Melophorus *Melophorus bagoti* Lubbock	Formicidae, Hymenoptera
Menacanthus *Menacanthus cornutus* (Schömmer) *Menacanthus eurysternus* (Burmeister) *Menacanthus pallidulus* (Neumann) *Menacanthus stramineus* (Nitzsch)	Menoponidae, Phthiraptera
Menopon *Menopon gallinae* (Linnaeus) *Menopon pallens* Clay *Menopon spiniferum* Piaget	Menoponidae, Phthiraptera **see** *Menacanthus eurysternus*

Meoneura *Meoneura amurensis* Ozerov	Carnidae, Diptera
Meringis *Meringis agilis* Eads *Meringis altipecten* Traub & Ioff *Meringis arachis* Jordan *Meringis bilsingi* Eads & Menzies *Meringis jamesoni* Hubbard *Meringis nidi* Williams & Hoff *Meringis parkeri* Jordan *Meringis rectus* Morlan *Meringis shannoni* (Jordan)	Hystrichopsyllidae, Siphonaptera
Meroplius *Meroplius minutus* (Wiedemann)	Sepsidae, Diptera
Mesalgoides *Mesalgoides dolichocaulus* Gaud *Mesalgoides mesocaulus* Gaud *Mesalgoides microcaulus* Gaud	Psoroptoididae, Acari
Mesembrina *Mesembrina resplendens* Wahlberg	Muscidae, Diptera
Mesobuthus *Mesobuthus eupeus* Koch *Mesobuthus tamulus* Fabricius	Buthidae, Scorpiones **see** *Buthotus tamulus*
Mesocyclops *Mesocyclops albicans* (Smith) *Mesocyclops aspericornis* Daday *Mesocyclops kieferi* Van de Velde *Mesocyclops leuckarti leuckarti* (Claus) *Mesocyclops leuckarti pilosus* Kiefer *Mesocyclops thermocyclopoides* Harada	Cyclopidae, Copepoda
Mesomyia *Mesomyia rubricornis* (Kröber)	Tabanidae, Diptera
Mesopangonius *Mesopangonius brackleyae* Burger *Mesopangonius philipi* Burger	Tabanidae, Diptera
Mesopsylla *Mesopsylla hebes* Jordan & Rothschild	Leptopsyllidae, Siphonaptera
Messor *Messor ebeninus* Santschi *Messor galla* (Mayr)	Formicidae, Hymenoptera
Metabinuncus *Metabinuncus asellisci* Uchikawa *Metabinuncus paracoelopos* Uchikawa	Myobiidae, Acari
Metacheyletia *Metacheyletia longisetosa* Atyeo, Kethley & Pérez *Metacheyletia obesa* Fain	Cheyletidae, Acari
Metacnephia *Metacnephia blanci* (Grenier & Theodorides) *Metacnephia nuragica* Rivosecchi, Raastad & Contini *Metacnephia uzunovi* Kovachev	Simuliidae, Diptera

Metacyclops
Metacyclops gracilis (Lilljeborg)
Metacyclops margaretae Lindberg
Metacyclops minutus (Claus)

Cyclopidae, Copepoda

see *Metacyclops gracilis*

Metallea
Metallea papua Kurahashi

Calliphoridae, Diptera

Metapenaeus
Metapenaeus monoceros (Fabricius)

Penaeidae, Decapoda

Metastivalius
Metastivalius novaehiberniae Beaucournu & Mahnert

Pygiopsyllidae, Siphonaptera

Metriocnemus
Metriocnemus fuscipes (Meigen)
Metriocnemus hygropetricus Kieffer
Metriocnemus knabi Coquillett
Metriocnemus obscuripes Holmgren
Metriocnemus yaquina Cranston & Judd

Chironomidae, Diptera

see *Metriocnemus obscuripes*

Michaelopus
Michaelopus johnstoni Fain
Michaelopus rwandanus Fain
Michaelopus tridens Fain & Lukoschus

Acaridae, Acari

Microargas

see *Argas*

Microcyclops
Microcyclops minutus (Claus)
Microcyclops varicans (Sars)

Cyclopidae, Copepoda
see *Metacyclops minutus*

Microlichus
Microlichus avus (Trouessart)
Microlichus chloris Fain, Gaud & Philips
Microlichus turdicola Fain, Gaud & Philips

Epidermoptidae, Acari

Micronychites
Micronychites postverrucosus Fain & Lukoschus

Rosensteiniidae, Acari

Microprosopa
Microprosopa haemorrhoidalis (Meigen)

Scathophagidae, Diptera

Micropsylla
Micropsylla sectilis (Jordan & Rothschild)

see *Rhadinopsylla*
see *Rhadinopsylla sectilis*

Microselia
Microselia southwoodi Disney

Phoridae, Diptera

Microthoracius
Microthoracius cameli (Linnaeus)
Microthoracius mazzai Werneck
Microthoracius minor Werneck
Microthoracius praelongiceps (Neumann)

Microthoraciidae, Phthiraptera

Microtityus
Microtityus dominicanensis Santiago–Blay

Buthidae, Scorpiones

Microtriatoma
Microtriatoma trinidadensis (Lent)

Reduviidae, Hemiptera

Microtrombicula
Microtrombicula balcanica Kolebinova
Microtrombicula brachytrichia Brennan

see *Ascoschoengastia*
see *Ascoschoengastia balcanica*
see *Ascoschoengastia brachytrichia*

Microtrombicula brennani Goff, Whitaker & Dietz	**see** *Ascoschoengastia brennani*
Microtrombicula hoplodactyla Goff, Loomis & Ainsworth	**see** *Ascoschoengastia hoplodactyla*
Microtrombicula nicaraguae Webb & Loomis	**see** *Ascoschoengastia nicaraguae*
Microvelia	Veliidae, Hemiptera
Microvelia pulchella Westwood	
Microvelia reticulata (Burmeister)	
Mimeustathia	Eustathiidae, Acari
Mimomyia	Culicidae, Diptera
Mimomyia hispida (Theobald)	
Mimomyia uniformis Theobald	**see** *Ficalbia uniformis*
Minosiella	Drassidae, Araneae
Minosiella intermedia Denis	
Mirobaeoides	Scelionidae, Hymenoptera
Misumena	Thomisidae, Araneae
Misumena vatia (Clerck)	
Mitchella	Ischnopsyllidae, Siphonaptera
Mitchella laxisinuata (Liu, Wu & Wu)	
Mitchella megatarsalia (Liu, Wu & Wu)	
Mitchella truncata (Liu, Wu & Wu)	
Miyatrombicula	Trombiculidae, Acari
Miyatrombicula benensoni Tanskul & Nadchatram	
Mochlonyx	Chaoboridae, Diptera
Mochlonyx cinctipes (Coquillett)	
Mochlonyx culiciformis (DeGeer)	**see** *Mochlonyx velutinus*
Mochlonyx velutinus (Ruthe)	
Moina	Moinidae, Branchiopoda
Moina brachiata (Jurine)	
Moina macrocopa (Straus)	
Moina micrura Kurz	
Molossilichus	Rosensteiniidae, Acari
Molossilichus macrobursatus Fain & Lukoschus	
Monapsidus	Avenzoariidae, Acari
Monapsidus cernyi Faccini & Atyeo	
Monapsidus cyrtotoxus Gaud	
Monapsidus tetrophtalmus Gaud	
Monobia	Eumenidae, Hymenoptera
Monobia quadridens (Linnaeus)	
Monohelea	Ceratopogonidae, Diptera
Monohelea baltica Szadziewski	
Monomorium	Formicidae, Hymenoptera
Monomorium bicolor Emery	
Monomorium bicolor nitidiventre Emery	**see** *Monomorium nitidiventre*
Monomorium destructor (Jerdon)	
Monomorium floricola (Jerdon)	
Monomorium minimum (Buckley)	
Monomorium nitidiventre Emery	
Monomorium pharaonis (Linnaeus)	

Monopis *Monopis congestella* (Walker) *Monopis impressella* (Walker) *Monopis longella* (Walker)	Tineidae, Lepidoptera
Monopsyllus *Monopsyllus anisus* (Rothschild) *Monopsyllus ciliatus* (Baker) *Monopsyllus eumolpi* (Rothschild) *Monopsyllus forficus* Cai & Wu *Monopsyllus hamutus* Cai & Wu *Monopsyllus indages* (Rothschild) *Monopsyllus paradoxus* (Scalon) *Monopsyllus sciurorum* (Schrank) *Monopsyllus wagneri* (Baker)	**see** *Ceratophyllus et al.* **see** *Ceratophyllus anisus* **see** *Ceratophyllus ciliatus* **see** *Eumolpianus eumolpi* **see** *Ceratophyllus forficus* **see** *Ceratophyllus hamutus* **see** *Ceratophyllus indages* **see** *Ceratophyllus paradoxus* **see** *Ceratophyllus sciurorum* **see** *Aetheca wagneri*
Monopylidium *Monopylidium crateriformis* (Goeze)	Dilepididae, Cyclophyllidea
Montisimulium *Montisimulium shevyakovi* (Dorogostaisky *et al.*)	**see** *Simulium* **see** *Simulium schevyakovi*
Morellia *Morellia asetosa* Baranov *Morellia hortensia* (Wiedemann) *Morellia hortorum* (Fallén) *Morellia saishuensis* Ôuchi *Morellia simplex* (Loew)	Muscidae, Diptera **see** *Morellia asetosa*
Mormoniella *Mormoniella vitripennis* (Walker)	**see** *Nasonia* **see** *Nasonia vitripennis*
Mortonagrion *Mortonagrion hirosei* Asahina	Coenagriidae, Odonata
Mothocya	**see** *Cymothon*
Multisetosa *Multisetosa sciurotamiatis* Zhou, Wen, Chen & Zhai *Multisetosa sylvatici* Zhou, Wen, Chen & Zhai	Leeuwenhoekiidae, Acari
Musca *Musca asiatica* Shinonaga & Kano *Musca autumnalis* DeGeer *Musca bakeri* Patton *Musca bezzii* Patton & Cragg *Musca conducens* Walker *Musca confiscata* Speiser *Musca craggi* Patton *Musca crassirostris* Stein *Musca domestica calleva* Walker *Musca domestica curviforceps* Saccà & Rivosecchi *Musca domestica domestica* Linnaeus *Musca domestica nebulo* Fabricius *Musca domestica vicina* Macquart *Musca hervei* Villeneuve *Musca inferior* Stein *Musca lasiophthalma* Thomson *Musca lusoria* Wiedemann *Musca nebulo* Fabricius *Musca nevilli* Kleynhans *Musca osiris* Wiedemann *Musca planiceps* Wiedemann *Musca sorbens* Wiedemann	Muscidae, Diptera **see** *Musca domestica domestica* **see** *Musca domestica domestica* **see** *Musca domestica domestica*

Musca tempestiva Fallén
Musca vetustissima Walker
Musca vitripennis Meigen
Musca xanthomelas Wiedemann

Muscidifurax   Pteromalidae, Hymenoptera
Muscidifurax raptor Girault & Sanders
Muscidifurax raptorellus Kogan & Legner
Muscidifurax uniraptor Kogan & Legner
Muscidifurax zaraptor Kogan & Legner

Muscina   Muscidae, Diptera
Muscina angustifrons Loew
Muscina assimilis (Fallén)   **see** *Muscina levida*
Muscina levida (Harris)
Muscina stabulans (Fallén)

Mutilla   Mutillidae, Hymenoptera

Mycotrupes   Scarabaeidae, Coleoptera

Mydaea   Muscidae, Diptera
Mydaea plaumanni Snyder

Myialges   Epidermoptidae, Acari
Myialges anchora Sergent & Trouessart
Myialges pari Fain
Myialges uncus (Vitzthum)

Mylabris   Meloidae, Coleoptera
Mylabris delhiensis Anand
Mylabris guptai Anand
Mylabris pustulata (Thunberg)

Myobia   Myobiidae, Acari
Myobia agraria Gorissen & Lukoschus
Myobia apodemi Uchikawa
Myobia kobayashii Uchikawa & Mizushima
Myobia multivaga Poppe
Myobia murismusculi Schrank   **see** *Myobia musculi*
Myobia musculi (Schrank)
Myobia nodae Matuzaki

Myocoptes   Myocoptidae, Acari
Myocoptes musculinus (Koch)
Myocoptes neotomae Fain *et al.*
Myocoptes queenslandicus Fain

Myodopsylla   Ischnopsyllidae, Siphonaptera
Myodopsylla insignis (Rothschild)

Myonyssus   Laelapidae, Acari
Myonyssus ingricus Bregetova
Myonyssus rossicus Bregetova

Myospila   Muscidae, Diptera
Myospila japonica Shinonaga & Iwasa
Myospila laevis (Stein)

Myoxopsylla   Ceratophyllidae, Siphonaptera
Myoxopsylla laverani (Rothschild)

Myrmecia   Formicidae, Hymenoptera
Myrmecia brevinoda Forel

Myrmecia gulosa (Fabricius)
Myrmecia pilosula F. Smith
Myrmecia rufinodis (F. Smith)
Myrmecia simillima F. Smith

Myrmecocystus   Formicidae, Hymenoptera
Myrmecocystus flaviceps Wheeler

Myrmica   Formicidae, Hymenoptera
Myrmica laevinodis Nylander
Myrmica ruginodis Nylander

Myrsidea   Menoponidae, Phthiraptera
Myrsidea cucullaris (Nitzsch)
Myrsidea serini (Séguy)

Mysolaelaps   Laelapidae, Acari
Mysolaelaps parvispinosus Fonseca

Mystacobia   Myobiidae, Acari

Myzopodobia   Myobiidae, Acari

N
Nabis   Nabidae, Hemiptera
Nabis capsiformis Germar
Nabis ferus (Linnaeus)

Nacerdes   Oedemeridae, Coleoptera
Nacerdes melanura (Linnaeus)

Nanacarus   Hemisarcoptidae, Acari
Nanacarus minutus (Oudemans)

Nanoptilium   Ptiliidae, Coleoptera
Nanoptilium aequisetum Młynarski
Nanoptilium flammiferum Młynarski

Nanorrhynchus   Tabanidae, Diptera

Naobranchia   Naobranchiidae, Copepoda

Nasicola   **see** *Brennanacarus*

Nasonia   Pteromalidae, Hymenoptera
Nasonia vitripennis (Walker)

Natalimyobia   Myobiidae, Acari

Naucoris   Naucoridae, Hemiptera
Naucoris cimicoides (Linnaeus)   **see** *Ilyocoris cimicoides*

Nauphoeta   Oxyhaloidae, Dictyoptera
Nauphoeta cinerea (Olivier)

Nearctopsylla   Hystrichopsyllidae, Siphonaptera
Nearctopsylla genalis (Baker)
Nearctopsylla liupanshanensis Li, Wu & Liu

Nebo   Diplocentridae, Scorpiones
Nebo hierichonticus (Simon)

Necrobia   Cleridae, Coleoptera
Necrobia ruficollis (Fabricius)

Necrobia rufipes (DeGeer)

Necrodes
Necrodes surinamensis (Fabricius)

Silphidae, Coleoptera

Necrophila
Necrophila americana (Linnaeus)

Silphidae, Coleoptera

Necrophorus
Necrophorus vespilloides Herbst

see *Nicrophorus*
see *Nicrophorus vespilloides*

Nemapalpus
Nemapalpus patriciae Alexander

Psychodidae, Diptera

Nemorius
Nemorius caucasicus (Olsuf'ev)

Tabanidae, Diptera

Neobrachiella
Neobrachiella anisotremi Romero & Kuroki
Neobrachiella robusta (Wilson)
Neobrachiella rostrata (Krøyer)

Lernaeopodidae, Copepoda

Neochauliacia
Neochauliacia decorata Gaud

Eustathiidae, Acari

Neocheyletiella
Neocheyletiella alfortensis Guilhon & Euzeby

Cheyletidae, Acari

Neocnemidocoptes
Neocnemidocoptes gallinae (Railliet)

Knemidokoptidae, Acari

Neocoptopsylla
Neocoptopsylla wassiliewi Wagner

see *Coptopsylla*
see *Coptopsylla wassiliewi*

Neocypholaelaps
Neocypholaelaps favus Ishikawa

Ameroseiidae, Acari

Neohaematopinus
Neohaematopinus rupestis Chin
Neohaematopinus sciuri Jancke
Neohaematopinus sciuropteri (Osborn)
Neohaematopinus setosus Chin

Polyplacidae, Phthiraptera

Neohelea
Neohelea pastoriana Clastrier

Ceratopogonidae, Diptera

Neoichoronyssus
Neoichoronyssus wernecki (Fonseca)

Macronyssidae, Acari

Neolipoptena
Neolipoptena ferrisi (Bequaert)

Hippoboscidae, Diptera

Neomuscina

Muscidae, Diptera

Neomyia
Neomyia cornicina (Fabricius)
Neomyia laevifrons Loew
Neomyia pacifica Zimin

Muscidae, Diptera

see *Neomyia laevifrons*

Neonyssus

Rhinonyssidae, Acari

Neopodocinum
Neopodocinum dehongense Li & Chang
Neopodocinum mrciaki Sellnick

Macrochelidae, Acari

Neopodocinum sinicum Li & Gu
Neopodocinum yunnanense Li & Gu

Neopsylla   Hystrichopsyllidae, Siphonaptera
Neopsylla abagaitui Ioff
Neopsylla bidentatiformis (Wagner)
Neopsylla setosa (Wagner)

Neoschoengastia   Trombiculidae, Acari
Neoschoengastia americana (Hirst)
Neoschoengastia israelensis Goff

Neoscona   Araneidae, Araneae
Neoscona minima F. Pickard-Cambridge

Neospeleognathopsis   Ereynetidae, Acari
Neospeleognathopsis molossus Fain & Lukoschus

Neostylopyga   Blattidae, Dictyoptera
Neostylopyga rhombifolia (Stoll)

Neosuidasia   Saproglyphidae, Acari
Neosuidasia faini Ranganath & ChannaBasavanna

Neotrombicula   Trombiculidae, Acari
Neotrombicula aeretes Hsu & Yang
Neotrombicula austriaca Kepka
Neotrombicula autumnalis (Shaw)
Neotrombicula earis Kepka
Neotrombicula gamaensis Yang
Neotrombicula ichikawai (Sasa)
Neotrombicula inopinata (Oudemans)
Neotrombicula japonica (Tanaka *et al.*)
Neotrombicula jiadingensis Yang
Neotrombicula kugitangica Amangulyev
Neotrombicula longichaeta Kudryashova & Abu-Taka
Neotrombicula mackayensis (Womersley)
Neotrombicula microti (Ewing)
Neotrombicula microtoides Hsu & Yang
Neotrombicula minuta Shluger
Neotrombicula mitamurai (Sasa *et al.*)
Neotrombicula nagayoi (Sasa *et al.*)
Neotrombicula naultini (Dumbleton)
Neotrombicula oeconomus Hsu & Yang
Neotrombicula pomeranzevi (Shluger)
Neotrombicula sphenodonti Goff, Loomis & Ainsworth
Neotrombicula tadjikistanica Kudryashova & Abu-Taka
Neotrombicula talmiensis (Shluger)
Neotrombicula tragardhiana (Feider)
Neotrombicula vulgaris (Shluger)
Neotrombicula zachvatkini (Shluger)   **see** *Hirsutiella zachvatkini*

Neottialges   Hypoderatidae, Acari
Neottialges pelagicus OConnor

Neottiophilum   Neottiphilidae, Diptera
Neottiophilum praeustum (Meigen)

Nepa   Nepidae, Hemiptera
Nepa chinensis Hoffmann
Nepa cinerea Linnaeus

Nephila   Argiopidae, Araneae
Nephila clavata L. Koch

Nephila maculata (Fabricius)
Nephila pilipes (Fabricius)

Nephotettix   Cicadellidae, Hemiptera
Nephotettix cincticeps (Uhler)

Nerocila   Cymothoidae, Isopoda
Nerocila acuminata Schiödte & Meinert
Nerocila depressa Milne-Edwards
Nerocila priacanthusi Kumari *et al.*
Nerocila serra Schiödte & Meinert

Nesolynx   Eulophidae, Hymenoptera
Nesolynx glossinae (Waterston)

Nesticodes   Theridiidae, Araneae
Nesticodes rufipes (Lucas)

Neurobezzia   Ceratopogonidae, Diptera

Neurotrixa   Muscidae, Diptera
Neurotrixa felsina (Walker)

Nicrophorus   Silphidae, Coleoptera
Nicrophorus defodiens Mannerheim
Nicrophorus humator (Gleditsch)
Nicrophorus investigator Zetterstedt
Nicrophorus orbicollis Say
Nicrophorus sayi Laporte
Nicrophorus tomentosus Weber
Nicrophorus vespilloides Herbst

Nigronia   Corydalidae, Megaloptera
Nigronia serricornis (Say)

Nigronirmus   Philopteridae, Phthiraptera
Nigronirmus densilimbus densilimbus (Nitzsch)
Nigronirmus densilimbus stadleri (Eichler)

Niptus   Ptinidae, Coleoptera
Niptus hololeucus (Faldermann)

Nitokra   Ameiridae, Copepoda
Nitokra sphaeromata Bowman

Nitzschiella   Goniodidae, Phthiraptera
Nitzschiella hilli (Bedford)

Nodele   Cheyletidae, Acari
Nodele simplex Wafa & Soliman

Nodonota   Chrysomelidae, Coleoptera
Nodonota puncticollis (Say)

Nosomma   Ixodidae, Acari
Nosomma monstrosum (Nuttall & Warburton)

Nosopsyllus   Ceratophyllidae, Siphonaptera
Nosopsyllus barbarus Jordan & Rothschild
Nosopsyllus consimilis (Wagner)
Nosopsyllus fasciatus (Bosc)
Nosopsyllus henleyi Rothschild
Nosopsyllus iranus Wagner & Argyropulo
Nosopsyllus laeviceps (Wagner)

Nosopsyllus mokrzeckyi (Wagner)

Notiopsylla   Pygiopsyllidae, Siphonaptera
Notiopsylla kerguelensis Taschenberg

Notochaeta   Sarcophagidae, Diptera
Notochaeta bufonivora Lopes & Vogelsang

Notocyphus   Pompilidae, Hymenoptera
Notocyphus dorsalis arizonicus Townes
Notocyphus dorsalis dorsalis Cresson

Notoedres   Sarcoptidae, Acari
Notoedres cati (Hering)
Notoedres cuniculi (Gerlach)
Notoedres dewitti Klompen *et al.*
Notoedres dohanyi Klompen *et al.*
Notoedres ismaili Klompen *et al.*
Notoedres muris (Mégnin)
Notoedres notoedres (Delafond & Bourguignon)   **see** *Notoedres cati*
Notoedres pahangi Klompen *et al.*
Notoedres tristis Fain & Marshall

Notonecta   Notonectidae, Hemiptera
Notonecta glauca Linnaeus
Notonecta hoffmanni Hungerford
Notonecta kirbyi Hungerford
Notonecta lunata Hungerford
Notonecta marmorea marmorea Fabricius
Notonecta marmorea viridis Delcourt
Notonecta petrunkevitchi Hutchinson
Notonecta uhleri Kirkaldy
Notonecta undulata Say
Notonecta unifasciata Guérin-Méneville
Notonecta viridis Delcourt   **see** *Notonecta marmorea viridis*

Nowickia   Tachinidae, Diptera
Nowickia nitida (Wulp)
Nowickia rostrata (Tothill)

Numidilipeurus   Philopteridae, Phthiraptera
Numidilipeurus lawrensis lawrensis (Bedford)
Numidilipeurus lawrensis tropicalis (Peters)
Numidilipeurus tropicalis (Peters)   **see** *Numidilipeurus lawrensis tropicalis*

Nuttalliella   Nuttalliellidae, Acari
Nuttalliella namaqua Bedford

Nycteribia   Nycteribiidae, Diptera
Nycteribia kolenatii Theodor & Moscona
Nycteribia pedicularia Latreille

Nycteridopsylla   Ischnopsyllidae, Siphonaptera
Nycteridopsylla ancyluris Jordan
Nycteridopsylla oligochaeta Rybin
Nycteridopsylla pentactena (Kolenati)

Nycteriglyphites   Rosensteiniidae, Acari
Nycteriglyphites panamensis Fain & Méndez

Nycteriglyphoides   Rosensteiniidae, Acari
Nycteriglyphoides delamarei Fain

Nycteriglyphus *Nycteriglyphus fuscus* Dood & Rockett *Nycteriglyphus longipilis* Fain & Lukoschus *Nycteriglyphus xeniariae* Fain & Lukoschus	Rosensteiniidae, Acari
Nycterimyobia	Myobiidae, Acari
Nycterophilia *Nycterophilia mormoopsis* Wenzel	Streblidae, Diptera

O

Obuchovia *Obuchovia auricoma* (Meigen)	**see** *Simulium* **see** *Simulium auricoma*
Ochetomyrmex *Ochetomyrmex auropunctatus* (Roger)	Formicidae, Hymenoptera
Odagmia *Odagmia aokii* Takahasi *Odagmia monticola* (Friederichs) *Odagmia nishijimai* Ono *Odagmia nitidifrons* (Edwards) *Odagmia ornata* (Meigen) *Odagmia pontina* (Rivosecchi) *Odagmia spinosa* (Doby & Deblock) *Odagmia variegata* (Meigen)	**see** *Simulium* **see** *Simulium aokii* **see** *Simulium monticola* **see** *Simulium nishijimai* **see** *Simulium nitidifrons* **see** *Simulium ornatum* **see** *Simulium pontinum* **see** *Simulium spinosum* **see** *Simulium variegatum*
Odontacarus *Odontacarus adelaidae* (Womersley) *Odontacarus australiensis* (Hirst) *Odontacarus majesticus* (Chen & Hsu) *Odontacarus tetrasetosus* Yu & Yang	Leeuwenhoekiidae, Acari
Odontopsyllus *Odontopsyllus quirosi* (Gil Collado)	Leptopsyllidae, Siphonaptera
Oeciacus *Oeciacus hirundinis* (Jenyns) *Oeciacus vicarius* Horváth	Cimicidae, Hemiptera
Oedemagena *Oedemagena tarandi* (Linnaeus)	**see** *Hypoderma* **see** *Hypoderma tarandi*
Oedemera *Oedemera sexualis* Marseul	Oedemeridae, Coleoptera
Oedemeronia *Oedemeronia sexualis* (Marseul)	**see** *Oedemera* **see** *Oedemera sexualis*
Oestroderma	Oestridae, Diptera
Oestromyia *Oestromyia leporina* (Pallas) *Oestromyia prodigiosa* Grunin	Hypodermatidae, Diptera
Oestrus *Oestrus aureoargentatus* Rodhain & Bequaert *Oestrus macdonaldi* Gedoelst *Oestrus ovis* Linnaeus *Oestrus variolosus* (Loew)	Oestridae, Diptera
Oiceoptoma *Oiceoptoma noveboracense* (Forster)	Silphidae, Coleoptera

Oiclus *Oiclus purvesii* (Becker)	Diplocentridae, Scorpiones
Okinawayusurika *Okinawayusurika otsurui* Sasa & Hasegawa	Chironomidae, Diptera
Olabidocarpus *Olabidocarpus miniopteri* Fain	Chirodiscidae, Acari
Olfersia *Olfersia coriacea* van der Wulp	Hippoboscidae, Diptera
Ommatoiulus *Ommatoiulus moreleti* (Lucas)	Blaniulidae, Diplopoda
Omosita *Omosita colon* (Linnaeus)	Nitidulidae, Coleoptera
Oncocladosoma *Oncocladosoma castaneum* (Attems)	Paradoxosomatidae, Diplopoda
Oncopeltus *Oncopeltus fasciatus* (Dallas)	Lygaeidae, Hemiptera
Onesia *Onesia fumicosta* Kurahashi *Onesia gilwea* Kurahashi *Onesia higae* Kurahashi *Onesia ismayi* Kurahashi *Onesia nakatae* Kurahashi *Onesia tibialis* (Macquart)	Calliphoridae, Diptera
Oniticellus *Oniticellus africanus* Harold *Oniticellus fulvus* (Goeze) *Oniticellus intermedius* (Reiche) *Oniticellus pallipes* (Fabricius) *Oniticellus spinipes* Roth	Scarabaeidae, Coleoptera **see** *Tiniocellus spinipes*
Onitis *Onitis alexis* Klug *Onitis aygulus* Latreille *Onitis belial* Fabricius *Onitis bordati* Cambefort *Onitis caffer* Boheman *Onitis crenatus* Reiche *Onitis deceptor* Péringuey *Onitis fulgidus* Klug *Onitis naviauxi* Cambefort *Onitis obscurus* Lansberge *Onitis pecuarius* Lansberge *Onitis perpunctatus* Balthasar *Onitis tortuosus* Houston *Onitis uncinatus* Klug *Onitis viridulus* Boheman	Scarabaeidae, Coleoptera **see** *Onitis alexis*
Ontholestes *Ontholestes murinus* (Linnaeus) *Ontholestes tessellatus* (Fourcroy)	Staphylinidae, Coleoptera
Onthophagus *Onthophagus alluvius* Howden & Cartwright *Onthophagus australis* Guérin-Méneville *Onthophagus bedeli* Reitter	Scarabaeidae, Coleoptera

Onthophagus belorhinus Bates
Onthophagus binodis (Thunberg)
Onthophagus bistiniocelloides Krikken
Onthophagus bonasus (Fabricius)
Onthophagus collinsi Krikken & Huijbregts
Onthophagus conterminus Petrovitz
Onthophagus dejongi Krikken
Onthophagus dorsipilulus Howden & Gill
Onthophagus drescheri Paulian
Onthophagus ferox Harold
Onthophagus foedus Boucomont
Onthophagus fracticornis (Preyssler)
Onthophagus furcatus (Fabricius)
Onthophagus gazella (Fabricius)
Onthophagus granulatus Boheman
Onthophagus grossepunctatus Reitter
Onthophagus haroldi Ballion
Onthophagus hecate (Panzer)
Onthophagus koryoensis Kim
Onthophagus lenzi Harold
Onthophagus minutus (Hausmann)
Onthophagus oklahomensis Brown
Onthophagus opacicollis Reitter
Onthophagus palatus Boucomont
Onthophagus parapalatus Krikken & Huijbregts
Onthophagus paroculus Krikken & Huijbregts
Onthophagus pennsylvanicus Harold
Onthophagus phillippsorum Krikken & Huijbregts
Onthophagus politus (Fabricius)
Onthophagus rotundicollis Lansberge
Onthophagus sagittarius (Fabricius)
Onthophagus setoculus Krikken & Huijbregts
Onthophagus sideki Krikken & Huijbregts
Onthophagus similis (Scriba)
Onthophagus taurus (Schreber)
Onthophagus unifasciatus (Schaller)
Onthophagus vacca (Linnaeus)
Onthophagus viduus Harold

Onthophilus   Histeridae, Coleoptera
Onthophilus kirni Ross

Onychiurus   Onychiuridae, Collembola
Onychiurus furcifera (Börner)
Onychiurus taimyricus Martynova

Ooencyrtus   Encyrtidae, Hymenoptera
Ooencyrtus johnsoni (Howard)
Ooencyrtus submetallicus (Howard)
Ooencyrtus trinidadensis J. C. Crawford
Ooencyrtus trinidadensis venatorius De Santis *et al.*   **see** *Ooencyrtus venatorius*
Ooencyrtus venatorius De Santis & Vidal Sarmiento

Oolathron   Aphelinidae, Hymenoptera
Oolathron mireyae De Santis

Ophiogongylus   Ixodorrhynchidae, Acari
Ophiogongylus breviscutum Lizaso
Ophiogongylus rotundus Lizaso

Ophionyssus   Dermanyssidae, Acari
Ophionyssus javanensis Micherdzinski & Lukoschus
Ophionyssus natricis (Gervais)

Ophthalmodex
Ophthalmodex juniatae Veal, Giesen & Whitaker

Demodicidae, Acari

Ophthalmopsylla
Ophthalmopsylla jettmari Jordan
Ophthalmopsylla praefecta (Jordan & Rothschild)
Ophthalmopsylla volgensis tuoliensis Yu, Ye & Liu
Ophthalmopsylla volgensis volgensis (Wagner & Ioff)

Leptopsyllidae, Siphonaptera

Ophyiulus
Ophyiulus verruculiger (Verhoeff)

Julidae, Diplopoda

Ophyra
Ophyra aenescens (Wiedemann)
Ophyra capensis (Wiedemann)
Ophyra hirtitibia (Stein)
Ophyra leucostoma (Wiedemann)
Ophyra solitaria Albuquerque

see *Hydrotaea*
see *Hydrotaea aenescens*
see *Hydrotaea capensis*
see *Hydrotaea hirtitibia*
see *Hydrotaea ignava*
see *Hydrotaea solitaria*

Opifex
Opifex fuscus Hutton

Culicidae, Diptera

Opisocrostis
Opisocrostis hirsutus (Baker)
Opisocrostis labis (Jordan & Rothschild)
Opisocrostis oregonensis Good & Prince
Opisocrostis tuberculatus cynomuris Jellison

Ceratophyllidae, Siphonaptera
see *Oropsylla hirsuta*
see *Oropsylla labis*
see *Oropsylla oregonensis*
see *Oropsylla tuberculata cynomuris*

Opisodasys
Opisodasys pseudarctomys (Baker)
Opisodasys vesperalis (Jordan)

Ceratophyllidae, Siphonaptera

Opisthacanthus

Ischnuridae, Scorpiones

Opisthoplatia
Opisthoplatia orientalis (Burmeister)

Blattidae, Dictyoptera

Oppiella
Oppiella neerlandica (Oudemans)

Oppiidae, Acari

Orchopeas
Orchopeas caedens caedens (Jordan)
Orchopeas caedens durus (Jordan)
Orchopeas dieteri (C. Fox)
Orchopeas howardii (Baker)
Orchopeas latens (Jordan)
Orchopeas leucopus (Baker)
Orchopeas nepos (Rothschild)
Orchopeas sexdentatus (Baker)

Ceratophyllidae, Siphonaptera

Oribatula

Oribatulidae, Acari

Ornithobius
Ornithobius bucephalus (Giebel)

Philopteridae, Phthiraptera

Ornithocheyletia
Ornithocheyletia lichmerae Smiley
Ornithocheyletia lonchurae Smiley

Cheyletidae, Acari

Ornithocoris
Ornithocoris pallidus Usinger
Ornithocoris toledoi Pinto

Cimicidae, Hemiptera

Ornithodoros	Argasidae, Acari
Ornithodoros amblus Chamberlin	
Ornithodoros capensis Neumann	
Ornithodoros concanensis Cooley & Kohls	
Ornithodoros coriaceus Koch	
Ornithodoros denmarki Kohls, Sonenshine & Clifford	
Ornithodoros dyeri Cooley & Kohls	
Ornithodoros erraticus erraticus (Lucas)	
Ornithodoros erraticus sonrai Sautet & Witkowski	
Ornithodoros faini Hoogstraal	
Ornithodoros gurneyi Warburton	
Ornithodoros kelleyi Cooley & Kohls	
Ornithodoros lahorensis Neumann	
Ornithodoros maritimus Vermeil & Marguet	
Ornithodoros moubata moubata (Murray)	
Ornithodoros moubata porcinus Walton	
Ornithodoros muesebecki Hoogstraal	
Ornithodoros papillipes (Birulya)	**see** *Ornithodoros tholozani*
Ornithodoros parkeri Cooley	
Ornithodoros peringueyi Bedford & Hewitt	
Ornithodoros piriformis Warburton	
Ornithodoros porcinus Walton	**see** *Ornithodoros moubata porcinus*
Ornithodoros puertoricensis Fox	
Ornithodoros savignyi (Audouin)	
Ornithodoros sonrai Sautet & Witkowski	**see** *Ornithodoros erraticus sonrai*
Ornithodoros spheniscus Hoogstraal *et al.*	
Ornithodoros tadaridae Černý & Dusbábek	
Ornithodoros tartakovskyi Olenev	
Ornithodoros tholozani (Laboulbène & Mégnin)	
Ornithodoros turicata americanus Marx	
Ornithodoros turicata turicata (Dugès)	
Ornithodoros yunkeri Keirans, Clifford & Hoogstraal	
Ornithoica	Hippoboscidae, Diptera
Ornithoica confluenta (Say)	
Ornithomya	Hippoboscidae, Diptera
Ornithomya anchineuria Speiser	**see** *Ornithomya fringillina*
Ornithomya avicularia (Linnaeus)	
Ornithomya chloropus Bergroth	
Ornithomya fringillina Curtis	
Ornithomyia	**see** *Ornithomya*
Ornithonyssus	Macronyssidae, Acari
Ornithonyssus bacoti (Hirst)	
Ornithonyssus bursa (Berlese)	
Ornithonyssus campester Micherdzinski & Domrow	
Ornithonyssus capensis Shepherd & Narro	
Ornithonyssus garridoi de la Cruz	
Ornithonyssus iheringi (Fonseca)	
Ornithonyssus kochi (Fonseca)	
Ornithonyssus latro Domrow	
Ornithonyssus noeli de la Cruz	
Ornithonyssus praedo Domrow	
Ornithonyssus roseinnesi (Zumpt & Till)	
Ornithonyssus stigmaticus Micherdzinski & Domrow	
Ornithonyssus sylviarum (Canestrini & Fanzago)	
Ornithonyssus taphozous Micherdzinski & Domrow	
Ornithopsylla	Pulicidae, Siphonaptera
Ornithopsylla laetitiae Rothschild	

Oropsylla	Ceratophyllidae, Siphonaptera
Oropsylla bacchi bacchi (Rothschild)	
Oropsylla bacchi johnsoni (Hubbard)	
Oropsylla bruneri (Baker)	
Oropsylla fota (Jordan)	
Oropsylla hirsuta (Baker)	
Oropsylla idahoensis (Baker)	
Oropsylla labis (Jordan & Rothschild)	
Oropsylla montana (Baker)	
Oropsylla oregonensis (Good & Prince)	
Oropsylla tuberculata cynomuris (Jellison)	
Oropsylla tuberculata ornata (Fox)	**see** *Oropsylla tuberculata cynomuris*
Oropsylla tuberculata tuberculata (Baker)	
Orthellia	**see** *Neomyia*
Orthellia caesarion (Meigen)	**see** *Neomyia cornicina*
Orthellia cornicina (Fabricius)	**see** *Neomyia cornicina*
Orthellia pacifica Zimin	**see** *Neomyia laevifrons*
Orthellia viridis (Wiedemann)	**see** *Neomyia cornicina*
Orthetrum	Libellulidae, Odonata
Orthetrum triangulare melania (Selys)	
Orthetrum triangulare triangulare (Selys)	
Orthochirus	Buthidae, Scorpiones
Orthochirus scrobiculosus (Grube)	
Orthocladius	Chironomidae, Diptera
Orthocladius consobrinus (Holmgren)	
Orthocladius fuscimanus (Kieffer)	
Orthohalarachne	Halarachnidae, Acari
Orthohalarachne attenuata (Banks)	
Orthopodomyia	Culicidae, Diptera
Orthopodomyia anopheloides (Giles)	
Orthopodomyia antanosyorum Rodhain & Boutonnier	
Orthopodomyia madecassorum Rodhain & Boutonnier	
Orthopodomyia milloti Doucet	
Orthopodomyia pulchripalpis (Rondani)	
Orthopodomyia signifera (Coquillett)	
Orthopodomyia vernoni van Someren	
Orycteroxenus	Glycyphagidae, Acari
Orycteroxenus soricis (Oudemans)	
Oryzotelphusa	**see** *Oziotelphusa*
Oryzotelphusa senex (Fabricius)	**see** *Oziotelphusa senex*
Otilipeurus	Philopteridae, Phthiraptera
Otilipeurus turmalis (Denny)	
Otobius	Argasidae, Acari
Otobius megnini (Dugès)	
Otodectes	Psoroptidae, Acari
Otodectes cynotis (Hering)	
Oxidus	Paradoxosomatidae, Diplopoda
Oxidus gracilis (Koch)	
Oxya	Acrididae, Orthoptera
Oxya chinensis (Thunberg)	
Oxya nitidula (Walker)	

Oxya velox chinensis (Thunberg)	**see** *Oxya chinensis*
Oxycopis *Oxycopis mcdonaldi* (Arnett) *Oxycopis thoracica* (Fabricius)	Oedemeridae, Coleoptera
Oxylipeurus *Oxylipeurus clavatus* McGregor *Oxylipeurus corpulentus* Clay *Oxylipeurus dentatus* (Sugimoto) *Oxylipeurus mesopelios colchicus* (Clay) *Oxylipeurus polytrapezius* (Nitzsch) *Oxylipeurus spangleri* Price & Emerson *Oxylipeurus tetraonis* Grube	Philopteridae, Phthiraptera
Oxysarcodexia *Oxysarcodexia confusa* Lopes *Oxysarcodexia diana* (Lopes) *Oxysarcodexia fluminensis* Lopes *Oxysarcodexia paulistanensis* (Mattos) *Oxysarcodexia thornax* (Walker) *Oxysarcodexia ventricosa* (Wulp)	Sarcophagidae, Diptera
Oxytrigona *Oxytrigona daemoniaca* Camargo *Oxytrigona mellicolor* (Packard)	**see** *Trigona* **see** *Trigona daemoniaca* **see** *Trigona mellicolor*
Oxyvinia *Oxyvinia excisa* (Lopes)	Sarcophagidae, Diptera
Oziotelphusa *Oziotelphusa senex* (Fabricius)	Parathelphusidae, Decapoda

P

Pachycondyla *Pachycondyla harpax* (Fabricius) *Pachycondyla sennaarensis* (Mayr)	Formicidae, Hymenoptera
Pachycrepoideus *Pachycrepoideus vindemmiae* (Rondani)	Pteromalidae, Hymenoptera
Pachylister *Pachylister caffer* (Erichson)	Histeridae, Coleoptera **see** *Pactolinus caffer*
Pactolinus *Pactolinus caffer* (Erichson)	Histeridae, Coleoptera
Paederus *Paederus australis* Guérin-Méneville *Paederus fuscipes* Curtis	Staphylinidae, Coleoptera
Paguristes *Paguristes frontalis* Milne-Edwards *Paguristes sulcatus* Baker	Paguridae, Decapoda
Palaemon *Palaemon paucidens* De Haan	Palaemonidae, Decapoda
Palaemonetes *Palaemonetes argentinus* Nobili *Palaemonetes pugio* Holthuis	Palaemonidae, Decapoda
Palaeopsylla *Palaeopsylla anserocepsoides* Zhang, Wu & Liu	Hystrichopsyllidae, Siphonaptera

Palaeopsylla brevifrontata Zhang, Wu & Liu
Palaeopsylla cisalpina Jordan & Rothschild
Palaeopsylla gromovi Argyropulo
Palaeopsylla soricis gromovi Argyropulo   **see** *Palaeopsylla gromovi*
Palaeopsylla soricis soricis (Dale)

Palamnaeus   Scorpionidae, Scorpiones
Palamnaeus bengalensis (Koch)   **see** *Heterometrus bengalensis*

Palexorista   Tachinidae, Diptera
Palexorista imberbis (Wiedemann)

Palmodes   Sphecidae, Hymenoptera
Palmodes occitanicus (Lepeletier & Serville)

Palomena   Pentatomidae, Hemiptera
Palomena prasina (Linnaeus)

Palpomyia   Ceratopogonidae, Diptera
Palpomyia flavipes (Meigen)

Paltothyreus   Formicidae, Hymenoptera
Paltothyreus tarsatus (Fabricius)

Palystes   Heteropodidae, Araneae
Palystes natalius (Karsch)

Pamphobeteus   Theraphosidae, Araneae

Pandinus   Scorpionidae, Scorpiones
Pandinus imperator (Koch)

Panesthia   Blaberidae, Dictyoptera
Panesthia angustipennis angustipennis (Illiger)
Panesthia angustipennis spadica Shiraki
Panesthia angustipennis yayeyamensis Asahina

Pangonius   Tabanidae, Diptera
Pangonius florae Leclercq & Maldès
Pangonius micans Meigen

Panstrongylus   Reduviidae, Hemiptera
Panstrongylus geniculatus (Latreille)
Panstrongylus herreri Wygodzinsky
Panstrongylus lignarius (Walker)
Panstrongylus megistus (Burmeister)
Panstrongylus rufotuberculatus (Champion)
Panstrongylus tupynambai Lent

Pantala   Libellulidae, Odonata
Pantala flavescens (Fabricius)
Pantala hymenaea (Say)

Parabelminus   Reduviidae, Hemiptera

Parabuthus   Buthidae, Scorpiones

Paraceroglyphus   Acaridae, Acari
Paraceroglyphus californicus Fain & Schwan
Paraceroglyphus cynomydis OConnor & Pfaffenberger

Parachartergus   Vespidae, Hymenoptera
Parachartergus aztecus Willink
Parachartergus fraternus (Gribodo)

Paracimex Paracimex borneensis Usinger	Cimicidae, Hemiptera
Paracyclops Paracyclops fimbriatus (Fischer)	Cyclopidae, Copepoda
Paradoxopsyllus Paradoxopsyllus calceiforma Zhang & Liu Paradoxopsyllus scalonae Violovich	Leptopsyllidae, Siphonaptera
Paraergasilus Paraergasilus markevichi Titar & Chernogorenko Paraergasilus rylovi Markewitsch	Ergasilidae, Copepoda see Paraergasilus rylovi
Paragnetina Paragnetina media (Walker)	Perlidae, Plecoptera
Paraguacarus Paraguacarus abrelli Goff & Whitaker Paraguacarus callosus Goff & Whitaker	Trombiculidae, Acari
Paraiurus Paraiurus nordmanni (Birula)	Iuridae, Scorpiones
Parakalumma Parakalumma lydia (Jacot)	Parakalumnidae, Acari
Parakosa Parakosa flexipes (Pinichpongse)	Chirodiscidae, Acari
Paraperiglischrus Paraperiglischrus rhinolophinus (Koch)	Spinturnicidae, Acari
Paraphauloppia	Oribatulidae, Acari
Paraphrissopoda Paraphrissopoda chrysostoma (Wiedemann)	Sarcophagidae, Diptera
Parapiophila Parapiophila vulgaris (Fallén)	Piophilidae, Diptera
Parapolybia Parapolybia indica (Saussure)	Vespidae, Hymenoptera
Parasarcophaga Parasarcophaga albiceps (Meigen) Parasarcophaga argyrostoma (Robineau-Desvoidy) Parasarcophaga assamensis Nandi & Ray Parasarcophaga barbata (Thomson) Parasarcophaga crassipalpis (Macquart) Parasarcophaga cyrtophorae Cantrell Parasarcophaga dux (Thomson) Parasarcophaga hirtipes (Wiedemann) Parasarcophaga knabi (Parker) Parasarcophaga macroauriculata (Ho) Parasarcophaga misera (Walker) Parasarcophaga oitana (Hori) Parasarcophaga orchidea (Böttcher) Parasarcophaga peregrina (Robineau-Desvoidy) Parasarcophaga ruficornis (Fabricius) Parasarcophaga scopariiformis (Senior White) Parasarcophaga uliginosa (Kramer)	see Sarcophaga see Sarcophaga albiceps see Sarcophaga argyrostoma see Sarcophaga assamensis see Sarcophaga argyrostoma see Sarcophaga crassipalpis see Sarcophaga cyrtophorae see Sarcophaga dux see Sarcophaga hirtipes see Sarcophaga sericea see Sarcophaga macroauriculata see Sarcophaga misera see Sarcophaga oitana see Sarcophaga misera see Sarcophaga peregrina see Sarcophaga ruficornis see Sarcophaga scopariiformis see Sarcophaga uliginosa

Parasecia *Parasecia gurneyi* (Ewing)	Trombiculidae, Acari
Parasitus *Parasitus berlesei* (Willmann) *Parasitus coleoptratorum* (Linnaeus) *Parasitus copridis* Costa *Parasitus fimetorum* (Berlese) *Parasitus loricatus* (Wankel) *Parasitus niveus* (Wankel)	Parasitidae, Acari **see** *Eugamasus berlesei* **see** *Parasitus loricatus*
Paratanytarsus *Paratanytarsus grimmii* (Schneider) *Paratanytarsus inopertus* (Walker)	Chironomidae, Diptera
Paratendipes	Chironomidae, Diptera
Paratrechina *Paratrechina longicornis* (Latreille)	Formicidae, Hymenoptera
Paratrichobius	Streblidae, Diptera
Paratrombicula *Paratrombicula enciscoensis* Goff & Whitaker	Trombiculidae, Acari
Paravespula *Paravespula germanica* (Fabricius) *Paravespula maculifrons* (Buysson) *Paravespula pensylvanica* (Saussure) *Paravespula vulgaris* (Linnaeus)	**see** *Vespula* **see** *Vespula germanica* **see** *Vespula maculifrons* **see** *Vespula pensylvanica* **see** *Vespula vulgaris*
Parcoblatta *Parcoblatta fulvescens* (Saussure & Zehntner) *Parcoblatta kyotensis* Asahina *Parcoblatta pennsylvanica* (DeGeer) *Parcoblatta virginica* (Brunner)	Blattellidae, Dictyoptera
Pardosa *Pardosa flavipalpis* F.Pickard-Cambridge *Pardosa lugubris* (Walckenaer) *Pardosa ramulosa* (McCook)	Lycosidae, Araneae
Paregle *Paregle cinerella* (Fallén)	Anthomyiidae, Diptera
Parthocoris	**see** *Reduvius*
Paruroctonus *Paruroctonus maritimus* Williams *Paruroctonus mesaensis* Stahnka	Vaejovidae, Scorpiones
Parvidens *Parvidens heischi* (Kirk & Lewis)	Psychodidae, Diptera
Passeroptes *Passeroptes dermicola* (Trouessart) *Passeroptes myrmecocichlae* Fain	Epidermoptidae, Acari
Pattonella *Pattonella intermutans* (Walker)	Sarcophagidae, Diptera
Peckia *Peckia abnormis* (Enderlein) *Peckia chrysostoma* (Wiedemann)	Sarcophagidae, Diptera **see** *Adiscochaeta abnormis* **see** *Paraphrissopoda chrysostoma*

Pectinibuthus	Buthidae, Scorpiones
Pectinibuthus birulai Fet	
Pectinopygus	Philopteridae, Phthiraptera
Pectinopygus gyricornis (Denny)	
Pedicinus	Pedicinidae, Phthiraptera
Pedicinus albidus Rudow	
Pedicinus eurygaster Burmeister	
Pedicinus obtusus Rudow	
Pediculus	Pediculidae, Phthiraptera
Pediculus capitis DeGeer	
Pediculus chapini Ewing	
Pediculus corporis DeGeer	**see** *Pediculus humanus*
Pediculus humanus Linnaeus	
Pediculus humanus capitis DeGeer	**see** *Pediculus capitis*
Pediculus humanus corporis DeGeer	**see** *Pediculus humanus*
Pegomya	Anthomyiidae, Diptera
Pegomya finitima Stein	
Pelecanectes	Hypoderidae, Acari
Pelecanectes apunctatus Pence & Courtney	
Pelecyoplus	Analgoidea, Acari
Pelecyoplus allocentrus Gaud	
Pelecyoplus aprosdocetus Gaud	
Pellonyssus	Macronyssidae, Acari
Peltodytes	Haliplidae, Coleoptera
Peltodytes edentulus (Le Conte)	
Peltogaster	Peltogastridae, Cirripedia
Peltogaster curvatus Kossmann	
Peltogaster paguri Rathke	
Peltogasterella	Peltogastridae, Cirripedia
Peltogasterella curvatus (Kossmann)	**see** *Peltogaster curvatus*
Peltogasterella paguri (Rathke)	**see** *Peltogaster paguri*
Penicillidia	Nycteribiidae, Diptera
Penicillidia dufourii Westwood	
Penicillidia monoceros Speiser	
Peniculus	Pennellidae, Copepoda
Peniculus asinus Kabata	
Pennella	Pennellidae, Copepoda
Pennella antarctica Quidor	**see** *Pennella balaenoptera*
Pennella anthonyi Quidor	**see** *Pennella balaenoptera*
Pennella balaenoptera Korea & Danielson	
Pennella cetti Quidor	**see** *Pennella balaenoptera*
Pennella charcoti Quidor	**see** *Pennella balaenoptera*
Pennella diodontis Oken	
Pennella filosa (Linnaeus)	
Pennella makaira Hogans	
Pennella sagitta Linnaeus	
Pennellus	**see** *Pennella*
Pentagonaspis	Trombiculidae, Acari
Pentagonaspis aravani Kudryashova & Rybin	

Pergamasus *Pergamasus brevicornis* (Berlese) *Pergamasus longicornis* (Berlese) *Pergamasus oxalis* Karg	Parasitidae, Acari
Pericoma *Pericoma chlifasica* Vaillant & Moubayed *Pericoma litanica* Vaillant & Moubayed	Psychodidae, Diptera
Periglischrus	Spinturnicidae, Acari
Perineus *Perineus circumfasciatus* Kéler *Perineus concinnoides* Kéler *Perineus concinnus* Kellogg & Chapman *Perineus macronecti* Palma & Pilgrim *Perineus nigrolimbatus* Giebel *Perineus oblongus* Kéler	Philopteridae, Phthiraptera

Periplaneta — Blattidae, Dictyoptera
Periplaneta americana (Linnaeus)
Periplaneta australasiae (Fabricius)
Periplaneta brunnea Burmeister
Periplaneta emarginata Karny — **see** *Periplaneta fuliginosa*
Periplaneta fuliginosa (Serville)
Periplaneta furcata Karny
Periplaneta japonica Karny
Periplaneta orientalis (Linnaeus) — **see** *Blatta orientalis*

Peripolipus — Podapolipidae, Acari
Peripolipus muraii Husband

Peripsocus — Peripsocidae, Psocoptera
Peripsocus stagnivagus Chapman

Peromyscopsylla — Leptopsyllidae, Siphonaptera
Peromyscopsylla fallax (Rothschild)
Peromyscopsylla himalaica australishaanxia Zhang & Liu
Peromyscopsylla himalaica himalaica (Rothschild)
Peromyscopsylla scaliforma Zhang & Liu
Peromyscopsylla silvatica (Meinert)

Petauralges — Psoroptidae, Acari
Petauralges mordax Domrow

Petrobia — Tetranychidae, Acari
Petrobia latens (O.F. Müller)

Petrorossia — Bombyliidae, Diptera
Petrorossia angustibasalis Hesse

Peucetia — Oxyopidae, Araneae
Peucetia viridans (Hentz)

Phaenicia — **see** *Lucilia*
Phaenicia coeruleiviridis (Macquart) — **see** *Lucilia coeruleiviridis*
Phaenicia cuprina (Wiedemann) — **see** *Lucilia cuprina*
Phaenicia eximia (Wiedemann) — **see** *Lucilia eximia*
Phaenicia ibis (Shannon) — **see** *Lucilia ibis*
Phaenicia mexicana (Macquart) — **see** *Lucilia mexicana*
Phaenicia ochricornis (Wiedemann) — **see** *Lucilia eximia*
Phaenicia sericata (Meigen) — **see** *Lucilia sericata*

Phaeotabanus	Tabanidae, Diptera
Phaeotabanus cajennensis (Fabricius)	
Phaeotabanus innotescens (Walker)	
Phanaeus	Scarabaeidae, Coleoptera
Phanaeus difformis Le Conte	
Phanaeus endymion Harold	
Phaonantho	Muscidae, Diptera
Phaonantho devia Albuquerque	
Phaonia	Muscidae, Diptera
Phaonia antennicrassa Xue	
Phaonia boleticola (Rondani)	
Phaonia fuscitibia Shinonaga & Kano	
Phaonia mimobitrigona Xue	
Phaonia nititerga Xue	
Phaonia ryukyuensis Shinonaga & Kano	
Phaonia vittifera (Zetterstedt)	
Pharyngomyia	Oestridae, Diptera
Pharyngomyia picta (Meigen)	
Pheidole	Formicidae, Hymenoptera
Pheidole dentata Mayr	
Pheidole fervens F. Smith	
Pheidole megacephala (Fabricius)	
Pheidole teneriffana Forel	
Pherbellia	Sciomyzidae, Diptera
Pherbellia austera Meigen	
Pherbellia cinerella (Fallén)	
Pherbellia fisheri Orth	
Pherbellia griseicollis (Becker)	
Pherbellia hackmani Rozkosny	
Pherbellia majuscula Rondani	**see** *Pherbellia austera*
Pherbellia ventralis (Fallén)	
Phidippus	Salticidae, Araneae
Philanthus	Sphecidae, Hymenoptera
Philanthus triangulum (Fabricius)	
Philichthyophaga	**see** *Pectinopygus*
Philichthyophaga gyricornis (Denny)	**see** *Pectinopygus gyricornis*
Philipomyia	Tabanidae, Diptera
Philipomyia aprica (Meigen)	
Philoliche	Tabanidae, Diptera
Philonthus	Staphylinidae, Coleoptera
Philonthus brunneus (Gravenhorst)	**see** *Philonthus sericans*
Philonthus cruentatus (Gmelin)	
Philonthus decorus (Gravenhorst)	
Philonthus discoideus (Gravenhorst)	
Philonthus flavolimbatus Erichson	
Philonthus hepaticus Erichson	
Philonthus lomatus Erichson	
Philonthus longicornis Stephens	
Philonthus nitidus (Fabricius)	
Philonthus sericans (Gravenhorst)	
Philonthus splendens (Fabricius)	
Philonthus subcingulatus Macleay	

Philonthus ventralis (Gravenhorst)

Philopterus   Philopteridae, Phthiraptera
Philopterus atratus Nitzsch
Philopterus corvi Linnaeus
Philopterus fringillae (Scopoli)
Philopterus sturni (Schrank)

Philornis   Anthomyiidae, Diptera

Phlebotomus   Psychodidae, Diptera
Phlebotomus alexandri Sinton
Phlebotomus andrejevi Shakirzyanova
Phlebotomus angustus Artemiev
Phlebotomus ansarii Lewis
Phlebotomus arabicus Theodor
Phlebotomus argentipes Annandale & Brunetti
Phlebotomus ariasi Tonnoir
Phlebotomus balcanicus Theodor
Phlebotomus bergeroti Parrot
Phlebotomus brevis Theodor & Mesghali
Phlebotomus caucasicus Martsinovsky
Phlebotomus chabaudi Croset, Abonnenc & Rioux
Phlebotomus chinensis Newstead
Phlebotomus chinensis longiductus Parrot   **see** *Phlebotomus longiductus*
Phlebotomus clydei Sinton   **see** *Sergentomyia clydei*
Phlebotomus colabaensis Young & Chalam
Phlebotomus dentatus Sinton   **see** *Sergentomyia dentata*
Phlebotomus duboscqi Neveu-Lemaire
Phlebotomus elgonensis Ngoka, Madel & Mutinga
Phlebotomus flaviscutellatus Mangabeira   **see** *Lutzomyia flaviscutellata*
Phlebotomus fortunatarum Ubeda Ontiveros *et al.*
Phlebotomus grekovi Khodukin   **see** *Sergentomyia grekovi*
Phlebotomus grimmi Porchinskiĭ
Phlebotomus heischi Kirk & Lewis   **see** *Parvidens heischi*
Phlebotomus intermedius Lutz & Neiva   **see** *Lutzomyia intermedia*
Phlebotomus kandelakii Shchurenkova
Phlebotomus kazeruni Theodor & Mesghali
Phlebotomus keshishiani Shchurenkova
Phlebotomus kiangsuensis Yao & Wu
Phlebotomus langeroni Nitzulescu
Phlebotomus longicuspis Nitzulescu
Phlebotomus longiductus Parrot
Phlebotomus longipalpis Lutz & Neiva   **see** *Lutzomyia longipalpis*
Phlebotomus longipes Parrot & Martin
Phlebotomus major major Annandale
Phlebotomus major neglectus Tonnoir
Phlebotomus major wui Yang & Xiong   **see** *Phlebotomus smirnovi*
Phlebotomus martini Parrot
Phlebotomus mascittii Grassi
Phlebotomus mongolensis Sinton
Phlebotomus naqbenius Lewis & Büttiker
Phlebotomus neglectus Tonnoir   **see** *Phlebotomus major neglectus*
Phlebotomus orientalis Parrot
Phlebotomus panamensis Shannon   **see** *Lutzomyia panamensis*
Phlebotomus papatasi (Scopoli)
Phlebotomus pedifer Lewis, Mutinga & Ashford
Phlebotomus perfiliewi perfiliewi Parrot
Phlebotomus perfiliewi transcaucasicus Perfil'ev
Phlebotomus perniciosus Newstead
Phlebotomus perniciosus tobbi Adler & Theodor **see** *Phlebotomus tobbi*
Phlebotomus rodhaini Parrot
Phlebotomus rossi De Meillon & Lavoipierre
Phlebotomus roubaudi Newstead   **see** *Phlebotomus duboscqi*

Phlebotomus saevus Parrot & Martin
Phlebotomus schwetzi Adler, Theodor & Parrot — see Sergentomyia schwetzi
Phlebotomus sergenti saevus Parrot & Martin — see Phlebotomus saevus
Phlebotomus sergenti sergenti Parrot
Phlebotomus sergenti similis Perfil'ev — see Phlebotomus similis
Phlebotomus sichuanensis Leng & Yin
Phlebotomus silvatica Raynal & Gaschen — see Sergentomyia silvatica
Phlebotomus simici Nitzulescu & Nitzulescu
Phlebotomus similis Perfil'ev
Phlebotomus smirnovi Perfil'ev
Phlebotomus squamipleuris Newstead — see Grassomyia squamipleuris
Phlebotomus tobbi Adler & Theodor
Phlebotomus transcaucasicus Perfil'ev — see Phlebotomus perfiliewi transcaucasicus
Phlebotomus tumenensis Wang & Chan
Phlebotomus turanicus Artem'ev
Phlebotomus vexator Coquillett — see Lutzomyia vexator
Phlebotomus wui Yang & Xiong — see Phlebotomus smirnovi
Phlebotomus yunshengensis Leng

Pholcus — Pholcidae, Araneae
Pholcus ancoralis L. Koch
Pholcus phalangioides (Fuesslin)

Pholeoixodes — see Ixodes
Pholeoixodes canisuga (Johnston) — see Ixodes canisuga

Phoneutria — Ctenidae, Araneae
Phoneutria nigriventer (Keyserling)

Phoniomyia — Culicidae, Diptera
Phoniomyia davisi Lane & Cerqueira
Phoniomyia deanei Lourenço-de-Oliveira
Phoniomyia pilicauda (Root)

Phoretodagmia — Simuliidae, Diptera
Phoretodagmia alajensis Rubtsov — see Simulium alajense
Phoretodagmia ephemerophila (Rubtsov) — see Simulium ephemerophilum
Phoretodagmia jani (Lewis) — see Simulium jani
Phoretodagmia obikumbensis Rubtsov — see Simulium ephemerophilum
Phoretodagmia rashidi (Lewis) — see Simulium rashidi
Phoretodagmia rithrogenophila (Konurbaev) — see Simulium rithrogenophilum

Phormia — Calliphoridae, Diptera
Phormia regina (Meigen)
Phormia terraenovae Robineau-Desvoidy

Phortica — see Amiota
Phortica variegata (Fallén) — see Amiota variegata

Phthiridium — Nycteribiidae, Diptera
Phthiridium biarticulatum Hermann
Phthiridium khabilovi Hůrka
Phthiridium phthisicum phthisicum (Speiser)
Phthiridium phthisicum transmotum Maa
Phthiridium simile Hůrka

Phumosia — Calliphoridae, Diptera
Phumosia viridis Kurahashi

Phygadeuon — Ichneumonidae, Hymenoptera
Phygadeuon yonedai Kusigemati

Phyllognathopus — Phyllognathopididae, Copepoda
Phyllognathopus viguieri (Maupas)

Physiphora *Physiphora aenea* (Fabricius) *Physiphora demandata* (Fabricius)	Otitidae, Diptera
Phytosarcophaga *Phytosarcophaga australis* Johnston & Tiegs	Sarcophagidae, Diptera **see** *Sarcophaga australis*
Picnoseus *Picnoseus nigropictus* (Denier)	Meloidae, Coleoptera
Pierretia *Pierretia baoxingensis* Feng & Ye *Pierretia calicifera* (Boettcher) *Pierretia granulata* (Kramer) *Pierretia nemoralis* (Kramer)	Sarcophagidae, Diptera **see** *Sarcophaga baoxingensis* **see** *Sarcophaga calicifera* **see** *Sarcophaga granulata* **see** *Sarcophaga nemoralis*
Piezosimulium *Piezosimulium jeanninae* Peterson	Simuliidae, Diptera
Pilogalumna *Pilogalumna tenuiclavus* (Berlese)	Oribatidae, Acari
Pintomyia *Pintomyia fischeri* (Pinto)	**see** *Lutzomyia* **see** *Lutzomyia fischeri*
Pionea *Pionea damastesalis* (Walker)	Pyralidae, Lepidoptera
Piophila *Piophila casei* (Linnaeus) *Piophila vulgaris* Fallén	Piophilidae, Diptera **see** *Parapiophila vulgaris*
Pisaura *Pisaura mirabilis* (Clerck)	Pisauridae, Araneae
Pison *Pison morosum* F. Smith *Pison spinolae* Shuckard	Sphecidae, Hymenoptera
Plagiognathus *Plagiognathus politus* Uhler	Miridae, Hemiptera
Platycentropus *Platycentropus radiatus* (Say)	Limnephilidae, Trichoptera
Platycleis *Platycleis tamerlana* (Saussure)	Tettigoniidae, Orthoptera
Platypodia *Platypodia granulosa* (Rüppell)	Xanthidae, Decapoda
Platypsyllus *Platypsyllus castoris* Ritsema	Leptinidae, Coleoptera
Platytropesa	Calliphoridae, Diptera
Plebeiogryllus *Plebeiogryllus guttiventris* (Walker)	Gryllidae, Orthoptera
Plecia *Plecia nearctica* Hardy	Bibionidae, Diptera
Pleopodias	Cymothoidae, Isopoda

Plesiophrictus *Plesiophrictus collinus* Pocock	Theraphosidae, Araneae
Pleurophorus *Pleurophorus caesus* (Creutzer) *Pleurophorus pannonicus* Petrovitz	Scarabaeidae, Coleoptera
Plocopsylla *Plocopsylla athena* Schramm & Lewis *Plocopsylla chiris* Jordan *Plocopsylla diana* Beaucournu, Gallardo & Launay *Plocopsylla kasogonaga* Schramm & Lewis *Plocopsylla kilya* Schramm & Lewis *Plocopsylla lewisi* Beaucournu & Gallardo *Plocopsylla nungui* Schramm & Lewis *Plocopsylla viracocha* Schramm & Lewis *Plocopsylla wolffsohni* Rothschild	Stephanocircidae, Siphonaptera
Plodia *Plodia interpunctella* (Hübner)	Pyralidae, Lepidoptera
Pneumocoptes *Pneumocoptes jellisoni* Baker *Pneumocoptes penrosei* (Weidman) *Pneumocoptes tiollaisi* Doby	Epidermoptidae, Acari
Pneumonyssoides *Pneumonyssoides caninum* (Chandler & Ruhe)	Halarachnidae, Acari
Pneumonyssus *Pneumonyssus africanus* Fain *Pneumonyssus caninum* Chandler & Ruhe *Pneumonyssus simicola* Banks	Halarachnidae, Acari **see** *Pneumonyssoides caninum*
Poecilochirus *Poecilochirus carabi* G. & R. Canestrini *Poecilochirus monospinosus* Wise, Hennessey & Axtell *Poecilochirus subterraneus* (Müller)	Parasitidae, Acari
Poecilus	**see** *Pterostichus*
Pogonocladius *Pogonocladius consobrinus* (Holmgren)	Chironomidae, Diptera **see** *Orthocladius consobrinus*
Pogonomyrmex *Pogonomyrmex badius* (Latreille) *Pogonomyrmex barbatus* (F. Smith) *Pogonomyrmex bicolor* Cole *Pogonomyrmex californicus* (Buckley) *Pogonomyrmex desertorum* Wheeler *Pogonomyrmex maricopa* Wheeler *Pogonomyrmex montanus* MacKay *Pogonomyrmex occidentalis* (Cresson) *Pogonomyrmex rugosus* Emery *Pogonomyrmex salinus* Olsen *Pogonomyrmex subnitidus* Emery	Formicidae, Hymenoptera
Polistes *Polistes annularis* (Linnaeus) *Polistes aurifer* Saussure *Polistes bernardii* Le Guillou *Polistes carolina* (Linnaeus) *Polistes chinensis antennalis* Pérez *Polistes chinensis chinensis* (Fabricius)	Vespidae, Hymenoptera **see** *Polistes fuscatus aurifer* **see** *Polistes stigma bernardii*

Polistes dominulus (Christ)
Polistes dubius Saussure   **see** Polistes stigma dubius
Polistes erythrocephalus Latreille
Polistes exclamans Viereck
Polistes fuscatus (Fabricius)
Polistes fuscatus aurifer Saussure
Polistes gallicus (Linnaeus)
Polistes hebraeus (Fabricius)   **see** Polistes olivaceus
Polistes infuscatus Lepeletier
Polistes instabilis Saussure
Polistes jadwigae Dalla Torre
Polistes maculipennis Saussure   **see** Polistes stigma maculipennis
Polistes manillensis Saussure   **see** Polistes stigma manillensis
Polistes medius Kojima
Polistes metricus Say
Polistes multipictus Smith   **see** Polistes stigma multipictus
Polistes novarae Saussure   **see** Polistes stigma novarae
Polistes occultus Kojima
Polistes olivaceus (DeGeer)
Polistes pacificus Fabricius
Polistes perplexus Cresson
Polistes riparius Yamane & Yamane
Polistes snelleni Saussure
Polistes stigma bernardii Le Guillou
Polistes stigma dubius Saussure
Polistes stigma maculipennis Saussure
Polistes stigma manillensis Saussure
Polistes stigma multipictus Smith
Polistes stigma novarae Saussure
Polistes stigma stigma (Fabricius)
Polistes stigma townsvillensis Soika
Polistes townsvillensis Soika   **see** Polistes stigma townsvillensis
Polistes watutus Kojima

Pollenia   Calliphoridae, Diptera
Pollenia alajensis Rohdendorf   **see** Pollenia dasypoda
Pollenia angustigena Wainwright
Pollenia dasypoda Portschinsky
Pollenia griseotomentosa Jacentkovský
Pollenia grunini Rognes
Pollenia hungarica Rognes
Pollenia intermedia Macquart
Pollenia longitheca Rognes
Pollenia luteovillosa Rognes
Pollenia mystica Rognes
Pollenia paragrunini Rognes
Pollenia pectinata Grunin
Pollenia pseudintermedia Rognes
Pollenia pseudorudis Rognes
Pollenia rudis (Fabricius)
Pollenia semicinerea Villeneuve

Polleniopsis   Calliphoridae, Diptera

Polybia   Vespidae, Hymenoptera
Polybia occidentalis (Olivier)
Polybia paulista von Ihering
Polybia scutellaris (White)
Polybia sericea (Olivier)

Polygenis   Rhopalopsyllidae, Siphonaptera
Polygenis bohlsi jordani (Costa Lima)
Polygenis klagesi (Rothschild)
Polygenis lakoi Guimarães

Polygenis nitidus Johnson
Polygenis tripus (Jordan)

Polypedilum   Chironomidae, Diptera
Polypedilum cultellatum Goetghebuer
Polypedilum halterale (Coquillett)
Polypedilum kyotoense (Tokunaga)
Polypedilum nubifer (Skuse)
Polypedilum pedatum excelsius Townes
Polypedilum pedatum pedatum Townes

Polyphaga   Polyphagidae, Dictyoptera
Polyphaga aegyptiaca (Linnaeus)
Polyphaga plancyi Bolivar

Polyplax   Polyplacidae, Phthiraptera
Polyplax borealis Ferris
Polyplax eriopepli Ewing
Polyplax humae Khan & Khan
Polyplax hurrianicus Mishra
Polyplax kondana Mishra
Polyplax phloeomydis Cuy
Polyplax reclinata (Nitzsch)
Polyplax serrata (Burmeister)
Polyplax sindensis Shafi, Samad & Rehana
Polyplax spinulosa (Burmeister)
Polyplax stephensi (Christophers & Newstead)
Polyplax wallacei Durden

Porocephalus   Porocephalidae, Pentastomida
Porocephalus armillatus (Wyman)
Porocephalus crotali Humboldt

Porthesia   **see** *Euproctis*

Portschinskia   Oestridae, Diptera

Portunus   Portunidae, Decapoda
Portunus pelagicus (Linnaeus)

Pothea   Reduviidae, Hemiptera

Probopyrus   Bopyridae, Isopoda
Probopyrus oviformis Nierstrasz & Brandis

Procambarus   Cambaridae, Decapoda
Procambarus clarkii (Girard)

Procaviopsylla   Pulicidae, Siphonaptera
Procaviopsylla creusae (Rothschild)

Procebalges   Psoroptidae, Acari

Procladius   Chironomidae, Diptera
Procladius bellus (Loew)
Procladius noctivagus (Kieffer)

Proctolaelaps   Ascidae, Acari
Proctolaelaps pygmaeus (Müller)
Proctolaelaps vandenbergi Ryke

Proctophyllodes   Proctophyllodidae, Acari
Proctophyllodes anthi Vitzthum
Proctophyllodes cotyledon Trouessart

Proctophyllodes danieli Cerny
Proctophyllodes glandarinus C.L. Koch
Proctophyllodes leucosticti Chirov & Mironov
Proctophyllodes macrophallus Cerny
Proctophyllodes poublani Gaud
Proctophyllodes puniceus Cerny
Proctophyllodes trisetosis Ewing & Stover

Prolepidoglyphus   Glycyphagidae, Acari
Prolepidoglyphus oregonensis Fain & Whitaker

Prolistrophorus   Listrophoridae, Acari

Propicimex   Cimicidae, Hemiptera
Propicimex tucmatiani (Wygodzinsky)

Prosevania   Evaniidae, Hymenoptera
Prosevania fuscipes (Illiger)
Prosevania punctata (Brullé)   **see** *Prosevania fuscipes*

Prosimulium   Simuliidae, Diptera
Prosimulium apoina Ono
Prosimulium daisetsense Uemoto, Okazawa & Onishi
Prosimulium decemarticulatum (Twinn)
Prosimulium dicum Dyar & Shannon
Prosimulium fontanum Syme & Davies
Prosimulium fuscum Syme & Davies
Prosimulium gibsoni (Twinn)
Prosimulium hirtipes (Fries)
Prosimulium jezonicum (Matsumura)
Prosimulium karibaense Ono
Prosimulium kiotoense Shiraki
Prosimulium magnum Dyar & Shannon
Prosimulium mixtum Syme & Davies
Prosimulium multicaulis Popov
Prosimulium onychodactylum Dyar & Shannon
Prosimulium petrosum Rubtsov
Prosimulium rufipes (Meigen)
Prosimulium tomosvaryi (Enderlein)
Prosimulium travisi Stone
Prosimulium yezoense Shiraki

Prospaltella   **see** *Encarsia*

Protocalliphora   Calliphoridae, Diptera
Protocalliphora avium Shannon & Dobrosky
Protocalliphora azurea (Fallén)
Protocalliphora hirudo Shannon & Dobroscky   **see** *Trypocalliphora braueri*
Protocalliphora maruyamensis Kano & Sinonaga

Protolichus   Pterolichidae, Acari
Protolichus lunula (Robin)

Protomyobia   Myobiidae, Acari
Protomyobia americana McDaniel
Protomyobia blarinae Lukoschus, Jeucken & Whitaker
Protomyobia onoi Jameson & Dusbábek
Protomyobia panamensis Lukoschus, Jeucken & Whitaker

Protonectarina   Vespidae, Hymenoptera
Protonectarina sylveirae (Saussure)

Protophormia	**see** *Phormia*
Protophormia terraenovae (Robineau–Desvoidy)	**see** *Phormia terraenovae*
Przhevalskiana	Oestridae, Diptera
Przhevalskiana aenigmatica Grunin	
Przhevalskiana silenus (Brauer)	
Psammodius	Scarabaeidae, Coleoptera
Psammodius gestroi Clouët	
Psammodius laevicollis Klug	
Psammodius sinicus Rakovic	
Psammodius thailandicus Balthasar	
Psammolestes	Reduviidae, Hemiptera
Psammolestes arthuri (Pinto)	
Pseudacteon	Phoridae, Diptera
Pseudacteon crawfordi Coquillett	
Pseudacteon obtusus Borgmeier	
Pseudalloptes	Pterolichidae, Acari
Pseudalloptes lambda Trouessart	**see** *Rhytidelasma lambda*
Pseudanuretes	Caligidae, Copepoda
Pseudanuretes papernai Kabata & Deets	
Pseudolernaeopodina	Lernaeopodidae, Copepoda
Pseudolernaeopodina synaphobranchi Hogans	
Pseudolynchia	Hippoboscidae, Diptera
Pseudolynchia canariensis (Macquart)	
Pseudolynchia garzettae (Rondani)	
Pseudomenopon	Menoponidae, Phthiraptera
Pseudomenopon pilosum Scopoli	
Pseudomyrmex	Formicidae, Hymenoptera
Pseudomyrmex ferrugineus (F. Smith)	
Pseudomyrmex triplarinus (Weddell)	
Pseudonirmus	Philopteridae, Phthiraptera
Pseudonirmus charcoti (Neumann)	
Pseudonirmus gurlti (Taschenberg)	
Pseudothelphusa	Pseudothelphusidae, Decapoda
Pseudothelphusa dilatata (Rathbun)	
Psilochorus	Pholcidae, Araneae
Psilochorus sphaeroides (L. Koch)	
Psilocnetha	**see** *Simulium*
Psilocnetha almae Yankovskiĭ & Kashkimbaev	**see** *Simulium almae*
Psilocnetha griseicollis (Becker)	**see** *Simulium griseicolle*
Psithyrus	Apidae, Hymenoptera
Psithyrus ashtoni (Cresson)	
Psorergates	Psorergatidae, Acari
Psorergates bos Johnston	**see** *Psorobia bos*
Psorergates cercopitheci Zumpt & Till	
Psorergates crocidurae Lukoschus	
Psorergates cryptotis Giesen & Lukoschus	
Psorergates muricola Fain	
Psorergates ovis Womersley	**see** *Psorobia ovis*

Psorergates rattus Fain & Goff
Psorergates urotrichi Giesen & Lukoschus

Psorobia   Psorergatidae, Acari
Psorobia bos (Johnston)
Psorobia elephantuli Giesen, Spicka & Whitaker
Psorobia lagomorphae Giesen, Spicka & Whitaker
Psorobia ovis (Womersley)

Psorophora   Culicidae, Diptera
Psorophora ciliata (Fabricius)
Psorophora cilipes (Fabricius)
Psorophora columbiae (Dyar & Knab)
Psorophora confinnis (Lynch Arribálzaga)
Psorophora cyanescens (Coquillett)
Psorophora dimidiata Cerqueira
Psorophora discolor (Coquillett)
Psorophora discrucians (Walker)
Psorophora ferox (Humboldt)
Psorophora howardii Coquillett
Psorophora pallescens Edwards
Psorophora signipennis (Coquillett)
Psorophora varinervis Edwards

Psoroptes   Psoroptidae, Acari
Psoroptes bovis (Gerlach)   **see** *Psoroptes ovis*
Psoroptes caprae Railliet   **see** *Psoroptes cuniculi*
Psoroptes communis bovis (Gerlach)   **see** *Psoroptes ovis*
Psoroptes communis communis (Fürstenberg)   **see** *Psoroptes equi*
Psoroptes communis cuniculi (Delafond)   **see** *Psoroptes cuniculi*
Psoroptes communis ovis (Hering)   **see** *Psoroptes ovis*
Psoroptes cuniculi (Delafond)
Psoroptes equi (Raspail)
Psoroptes natalensis Hirst
Psoroptes ovis (Hering)

Psychoda   Psychodidae, Diptera
Psychoda alternata Say
Psychoda cinerea Banks
Psychoda parthenogenetica Tonnoir
Psychoda severini Tonnoir

Psychodopygus   **see** *Lutzomyia*
Psychodopygus ayrozai (Barretto & Coutinho)   **see** *Lutzomyia ayrozai*
Psychodopygus carrerai (Barretto)   **see** *Lutzomyia carrerai*
Psychodopygus chagasi (Costa Lima)   **see** *Lutzomyia chagasi*
Psychodopygus claustrei (Abonnenc, Léger & Fauran) **see** *Lutzomyia claustrei*
Psychodopygus complexus (Mangabeira)   **see** *Lutzomyia complexa*
Psychodopygus davisi (Root)   **see** *Lutzomyia davisi*
Psychodopygus hirsutus (Mangabeira)   **see** *Lutzomyia hirsuta*
Psychodopygus intermedius (Lutz & Neiva)   **see** *Lutzomyia intermedia*
Psychodopygus leonidasdeanei Fraiha et al.   **see** *Lutzomyia leonidasdeanei*
Psychodopygus llanosmartinsi Fraiha & Ward   **see** *Lutzomyia llanosmartinsi*
Psychodopygus paraensis (Costa Lima)   **see** *Lutzomyia paraensis*
Psychodopygus squamiventris maripaensis (Floch & Abonnenc) **see** *Lutzomyia squamiventris maripaensis*
Psychodopygus squamiventris squamiventris (Lutz & Neiva) **see** *Lutzomyia squamiventris squamiventris*
Psychodopygus wellcomei Fraiha, Shaw & Lainson   **see** *Lutzomyia wellcomei*
Psychodopygus yucumensis Le Pont et al.   **see** *Lutzomyia yucumensis*

Psylloglyphus   Saproglyphidae, Acari
Psylloglyphus crenulatus Fain & Beaucournu
Psylloglyphus micronychus Fain & Beaucournu

Pteracarus *Pteracarus faini* Uchikawa *Pteracarus miniopteri* Uchikawa	Myobiidae, Acari
Pterolichus *Pterolichus dilatatus* Trouessart *Pterolichus martini* Trouessart *Pterolichus tenuis* Trouessart *Pterolichus tritiventris* Trouessart	Pterolichidae, Acari **see** *Rhytidelasma dilatata* **see** *Laronyssus martini* **see** *Rhytidelasma tenuis* **see** *Rhytidelasma tritiventris*
Pteronarcella	Pteronarcidae, Plecoptera
Pteronyssoides *Pteronyssoides pastoris* Mironov *Pteronyssoides yungipicinus* Mironov	Avenzoariidae, Acari
Pteronyssus	Epidermoptidae, Acari
Pterophagoides *Pterophagoides bathmourus* Gaud & Mouchet *Pterophagoides paradoxus* Gaud & Barré *Pterophagoides santanai* Gaud & Barré *Pterophagoides talpacoti* Černý	Dermoglyphidae, Acari **see** *Byersalges talpacoti*
Pterostichus *Pterostichus aethiops* Panzer *Pterostichus melanarius* (Illiger) *Pterostichus niger* (Schaller)	Carabidae, Coleoptera
Pthirus *Pthirus pubis* (Linnaeus)	Pthiridae, Phthiraptera
Ptilonyssus *Ptilonyssus pentagonicus* Fain & Lukoschus *Ptilonyssus philemoni* Domrow	Rhinonyssidae, Acari **see** *Ptilonyssus philemoni*
Ptilopsaltis *Ptilopsaltis santarosae* Davis	Tineidae, Lepidoptera
Ptychophallus *Ptychophallus richmondi* (Rathbun) *Ptychophallus tumimanus* Rathbun	Pseudothelphusidae, Decapoda
Pulaeus *Pulaeus zaherii* El-Bishlawy & Rahka	Cunaxidae, Acari
Pulex *Pulex irritans* Linnaeus *Pulex simulans* Baker	Pulicidae, Siphonaptera
Pulicella *Pulicella aenigma* Lewis & Cheetham	Pulicidae, Siphonaptera
Punctoribates *Punctoribates manzanoensis* Hammer	Mycobatidae, Acari
Pycnoscelus *Pycnoscelus surinamensis* (Linnaeus)	Oxyhaloidae, Dictyoptera
Pyemotes *Pyemotes herfsi* (Oudemans) *Pyemotes tritici* (Lagrèze-Fossat & Montagné) *Pyemotes ventricosus* (Newport)	Pyemotidae, Acari

Pygmephorus   Pygmephoridae, Acari
Pygmephorus brevipes Savulkina
Pygmephorus erlangensis Krczal
Pygmephorus forcipatus Willmann
Pygmephorus iglehartae Smiley & Whitaker
Pygmephorus islandicus Sellnick
Pygmephorus lukoschusi Smiley & Whitaker
Pygmephorus russellae Smiley & Whitaker
Pygmephorus stammeri Krczal
Pygmephorus sylvilagus Kaliszewski & Rack
Pygmephorus utmarae Smiley & Whitaker
Pygmephorus whitakeri Mahunka
Pygmephorus wilsoni Smiley & Whitaker

Pyractomena   Lampyridae, Coleoptera
Pyractomena lucifera Melsheimer

Pyrellia   Muscidae, Diptera
Pyrellia aenea (Zetterstedt)   see Pyrellia rapax
Pyrellia rapax (Harris)
Pyrellia tasmaniae Macquart
Pyrellia tateyamensis Shinonaga

Pyroglyphus   Pyroglyphidae, Acari
Pyroglyphus africanus (Hughes)   see Hughesiella africana

Pyrota   Meloidae, Coleoptera
Pyrota deceptiva Selander

Q
Qinghailaeiaps   Laelapidae, Acari
Qinghailaelaps marmotae Gu & Yang

Quadraceps   Philopteridae, Phthiraptera
Quadraceps obliquus Mjöberg
Quadraceps ornatus Grube

R
Radfordia   Myobiidae, Acari
Radfordia aethomys aethomys Curfs, Lukoschus & Fain
Radfordia aethomys chrysophila Curfs, Lukoschus & Fain
Radfordia affinis (Poppe)
Radfordia daltoni Scheperboer, Lukoschus & Fain
Radfordia ensifera (Poppe)
Radfordia holochilus Lukoschus & Cock

Radfordiana   Trombiculidae, Acari
Radfordiana muri Estrada-Peña et al.

Raillietia   Laelapidae, Acari
Raillietia auris (Leidy)
Raillietia caprae Quintero, Bassols & Acevedo
Raillietia manfredi Domrow

Raillietiella   Cephalobaenidae, Pentastomida
Raillietiella affinis Bovien
Raillietiella hemidactyli Hett
Raillietiella teagueselfi Riley, McAllister & Freed   see Raillietiella hemidactyli

Rallicola   Philopteridae, Phthiraptera
Rallicola irediparrae Price & Emerson

Ranatra   Nepidae, Hemiptera
Ranatra dispar Montandon

Ranatra elongata Fabricius
Ranatra filiformis (Fabricius)
Ranatra linearis (Linnaeus)

Raphignathus   Raphignathidae, Acari

Rasahus   Reduviidae, Hemiptera
Rasahus costaricensis Coscaron & Maldonado-Capriles

Ravinia   Sarcophagidae, Diptera
Ravinia belforti (Prado & Fonseca)
Ravinia derelicta (Walker)
Ravinia lherminieri (Robineau-Desvoidy)
Ravinia pernix (Harris)
Ravinia querula (Walker)
Ravinia striata (Fabricius)   **see** Ravinia pernix
Ravinia ventricosa (Wulp)   **see** Oxysarcodexia ventricosa

Raymondia   Streblidae, Diptera
Raymondia huberi Frauenfeld

Reduviolus   **see** Nabis

Reduvius   Reduviidae, Hemiptera
Reduvius christophi (Jakowleff)
Reduvius dicki Carayon
Reduvius personatus (Linnaeus)

Reighardia   Linguatulidae, Acari

Renocila   Cymothoidae, Isopoda
Renocila bollandi Williams & Williams
Renocila kohnoi Williams & Williams
Renocila yamazatoi Williams & Williams

Reticulipeurus   Philopteridae, Phthiraptera
Reticulipeurus colchicus (Clay)   **see** Oxylipeurus mesopelios colchicus
Reticulipeurus mesopelios colchicus (Clay)   **see** Oxylipeurus mesopelios colchicus
Reticulipeurus tetraonis Grube   **see** Oxylipeurus tetraonis

Rhadinopsylla   Hystrichopsyllidae, Siphonaptera
Rhadinopsylla biconcava Chen, Ji & Wu
Rhadinopsylla caucasica Argyropulo
Rhadinopsylla mesoides Smit
Rhadinopsylla pentacantha (Rothschild)
Rhadinopsylla sectilis Jordan & Rothschild
Rhadinopsylla stenofronta Xie
Rhadinopsylla ucrainica Wagner & Argyropulo
Rhadinopsylla ukrainica Wagner & Argyropulo   **see** Rhadinopsylla ucrainica

Rhantus   Dytiscidae, Coleoptera
Rhantus bistriatus Bergstraesser
Rhantus consputus (Sturm)
Rhantus suturellus (Harris)   **see** Rhantus bistriatus

Rhechostica   Theraphosidae, Araneae
Rhechostica chalcodes (Chamberlin)
Rhechostica echina (Chamberlin)
Rhechostica seemanni (F.O. Pickard-Cambridge)

Rhinia   Calliphoridae, Diptera

Rhinocoris   **see** Rhynocoris

Rhinoestrus
Rhinoestrus latifrons Gan
Rhinoestrus purpureus (Brauer)
Rhinoestrus usbekistanicus Gan

Oestridae, Diptera

Rhinolophopsylla
Rhinolophopsylla unipectinata turkestanica Ioff
Rhinolophopsylla unipectinata unipectinata (Taschenberg)

Ischnopsyllidae, Siphonaptera

Rhinonyssus
Rhinonyssus himantopus Strandtmann

Rhinonyssidae, Acari

Rhipicentor
Rhipicentor nuttalli Cooper & Robinson

Ixodidae, Acari

Rhipicephalus
Rhipicephalus appendiculatus Neumann
Rhipicephalus arnoldi Theiler & Zumpt
Rhipicephalus bequaerti Zumpt
Rhipicephalus bergeoni Morel & Balis
Rhipicephalus bursa Canestrini & Fanzago
Rhipicephalus camicasi Morel, Mouchet & Rodhain
Rhipicephalus capensis Koch
Rhipicephalus compositus Neumann
Rhipicephalus cuspidatus Neumann
Rhipicephalus distinctus Bedford
Rhipicephalus evertsi evertsi Neumann
Rhipicephalus evertsi mimeticus Dönitz
Rhipicephalus follis Dönitz
Rhipicephalus glabroscutatum du Toit
Rhipicephalus guilhoni Morel & Vassiliades
Rhipicephalus haemaphysaloides Supino
Rhipicephalus humeralis Rondelli
Rhipicephalus hurti Wilson
Rhipicephalus jeanneli Neumann
Rhipicephalus kochi Dönitz
Rhipicephalus leporis Pomerantsev
Rhipicephalus lunulatus Neumann
Rhipicephalus muhsamae Morel & Vassiliades
Rhipicephalus nitens Neumann
Rhipicephalus oculatus Neumann
Rhipicephalus praetextatus Gerstaecker
Rhipicephalus pravus Dönitz
Rhipicephalus pulchellus (Gerstaecker)
Rhipicephalus pumilio Schulze
Rhipicephalus punctatus Warburton
Rhipicephalus pusillus Gil Collado
Rhipicephalus reichenowi Zumpt
Rhipicephalus rossicus Yakimov & Kohl-Yakimova
Rhipicephalus sanguineus (Latreille)
Rhipicephalus schulzei Olenev
Rhipicephalus secundus Feldman-Muhsam
Rhipicephalus senegalensis Koch
Rhipicephalus simpsoni Nuttall
Rhipicephalus simus senegalensis Koch
Rhipicephalus simus simus Koch
Rhipicephalus sulcatus Neumann
Rhipicephalus tetracornus Kitaoka & Suzuki
Rhipicephalus tricuspis Dönitz
Rhipicephalus turanicus Pomerantsev
Rhipicephalus zambeziensis Walker, Norval & Corwin

Ixodidae, Acari

see *Rhipicephalus turanicus*

see *Rhipicephalus senegalensis*

Rhodnius	Reduviidae, Hemiptera
Rhodnius ecuadoriensis Lent & León	
Rhodnius nasutus Stål	
Rhodnius neglectus Lent	
Rhodnius neivai Lent	
Rhodnius pallescens Barber	
Rhodnius pictipes Stål	
Rhodnius prolixus Stål	
Rhodnius robustus Larrousse	
Rhodothemis	Libellulidae, Odonata
Rhodothemis rufa (Rambur)	
Rhopalopsyllus	Rhopalopsyllidae, Siphonaptera
Rhopalurus	Buthidae, Scorpiones
Rhoptromeris	Eucoilidae, Hymenoptera
Rhoptromeris heptoma Hartig	
Rhynchosciara	Sciaridae, Diptera
Rhynchosciara americana (Wiedemann)	
Rhyncophoromyia	Phoridae, Diptera
Rhyncophoromyia conica (Malloch)	
Rhynocoris	Reduviidae, Hemiptera
Rhynocoris iracundus (Poda)	
Rhyopsocus	Psoquillidae, Psocoptera
Rhyopsocus phillipsae Sommerman	
Rhyparobia	Oxyhaloidae, Dictyoptera
Rhyparobia maderae (Fabricius)	
Rhytidelasma	Pterolichidae, Acari
Rhytidelasma bicostata Atyeo & Pérez	
Rhytidelasma cornigera Atyeo & Pérez	
Rhytidelasma dilatata (Trouessart)	
Rhytidelasma lambda (Trouessart)	
Rhytidelasma mesomexicana Atyeo, Gaud & Pérez	
Rhytidelasma tenuis (Trouessart)	
Rhytidelasma tritiventris (Trouessart)	
Rhytidelasma ulocerca (Trouessart)	
Rhytidelasma urophila Atyeo & Pérez	
Rhytidoponera	Formicidae, Hymenoptera
Rhytidoponera metallica (F. Smith)	
Riccardoella	Ereynetidae, Acari
Riccardoella limacum (Schrank)	
Rivellia	Platystomatidae, Diptera
Rivellia quadrifasciata (Macquart)	
Rocinela	Aegidae, Isopoda
Rocinela kapala Bruce	
Rocinela oculata (Harger)	
Rodhainyssus	Gastronyssidae, Acari
Rodhainyssus longipilis Fain	
Ropalidia	Vespidae, Hymenoptera
Ropalidia fasciata (Fabricius)	

Ropalidia formosa (Saussure)
Ropalidia marginata (Lepeletier)
Ropalidia revolutionalis (Saussure)

Rudnicula   Trombiculidae, Acari
Rudnicula dimolinae (Audy)
Rudnicula knighti (Radford)

Rudnicula leytensis Brown, Goff & Nadchatram
Rudnicula meilingensis Zhao

Runchomyia   Culicidae, Diptera
Runchomyia frontosa Theobald

Rypellia   Muscidae, Diptera

S
Sabethes   Culicidae, Diptera
Sabethes chloropterus (Humboldt)
Sabethes cyaneus (Fabricius)
Sabethes quasicyaneus Peryassú
Sabethes undosus (Coquillett)

Sacculina   Sacculinidae, Cirripedia
Sacculina granifera Boschma

Saemundssonia   Philopteridae, Phthiraptera
Saemundssonia stresemanni Timmermann

Saimirioptes   Audycoptidae, Acari
Saimirioptes hershkovitzi OConnor

Salganea   Blaberidae, Dictyoptera
Salganea esakii Roth
Salganea taiwanensis ryukyuanus Asahina
Salganea taiwanensis taiwanensis Roth

Salmincola   Lernaeopodidae, Copepoda
Salmincola extensus (Kessler)
Salmincola salmoneus (Linnaeus)

Saltella   Sepsidae, Diptera
Saltella nigripes Robineau-Desvoidy
Saltella orientalis Hendel
Saltella sphondylii (Schrank)

Salticus   Salticidae, Araneae
Salticus scenicus (Clerck)

Sancassania   **see** *Caloglyphus*
Sancassania berlesei (Michael)   **see** *Caloglyphus berlesei*

Saprinus   Histeridae, Coleoptera
Saprinus speciosus Erichson   **see** *Saprinus splendens*
Saprinus splendens (Paykull)

Sarcodexia   Sarcophagidae, Diptera
Sarcodexia sternodontis Townsend

Sarconesia   Calliphoridae, Diptera
Sarconesia chlorogaster (Wiedemann)
Sarconesia versicolor Bigot

Sarcophaga   Sarcophagidae, Diptera
Sarcophaga albiceps Meigen
Sarcophaga allisoni Sugiyama, Shinonaga & Kano
Sarcophaga aratrix Pandellé
Sarcophaga argyrostoma (Robineau–Desvoidy)
Sarcophaga assamensis (Nandi & Ray)
Sarcophaga australis (Johnston & Tiegs)
Sarcophaga baoxingensis (Feng & Ye)
Sarcophaga barbata Thomson   **see** *Sarcophaga argyrostoma*
Sarcophaga baruai Sugiyama
Sarcophaga birganjensis Sugiyama
Sarcophaga brevicornis Ho
Sarcophaga bullata Parker   **see** *Neobellieria bullata*
Sarcophaga calicifera Boettcher
Sarcophaga carnaria (Linnaeus)
Sarcophaga claviger Blackith & Blackith
Sarcophaga communis (Parker)   **see** *Ravinia querula*
Sarcophaga crassipalpis Macquart
Sarcophaga cruentata Meigen
Sarcophaga cyrtophorae (Cantrell)
Sarcophaga disneyi Blackith & Blackith
Sarcophaga dumoga Sugiyama & Kurahashi
Sarcophaga dux Thomson
Sarcophaga falculata Pandellé
Sarcophaga forceps Blackith & Blackith
Sarcophaga formosensis (Kirner & Lopes)
Sarcophaga gorokaensis Sugiyama, Shinonaga & Kano
Sarcophaga granulata Kramer
Sarcophaga haemorrhoidalis (Fallén)   **see** *Sarcophaga cruentata*
Sarcophaga hirtipes Wiedemann
Sarcophaga horti Blackith & Blackith
Sarcophaga imitatrix (Lopes)
Sarcophaga indusa (Curran)
Sarcophaga invaria Walker
Sarcophaga javanica (Lopes)
Sarcophaga karachiensis Bilqees, Ali & Khan
Sarcophaga karnyi Hardy
Sarcophaga knabi Parker   **see** *Sarcophaga sericea*
Sarcophaga koimani (Kano & Shinonaga)
Sarcophaga macroauriculata Ho
Sarcophaga martellata Senior White
Sarcophaga melanura Meigen
Sarcophaga misera Walker
Sarcophaga nathani Lopes
Sarcophaga nemoralis Kramer
Sarcophaga nepalensis Kano & Shinonaga
Sarcophaga nodosa Engel
Sarcophaga oitana Hori
Sarcophaga orchidea Böttcher   **see** *Sarcophaga misera*
Sarcophaga pattoni Senior White
Sarcophaga peregrina (Robineau–Desvoidy)
Sarcophaga ruficornis (Fabricius)
Sarcophaga scoparia Pandellé
Sarcophaga scopariiformis Senior White
Sarcophaga septentrionalis (Rodendorf)
Sarcophaga sericea Walker
Sarcophaga shresthai Kano & Shinonaga
Sarcophaga situliformis (Zhong, Wu & Fan)
Sarcophaga subvicina Rodendorf
Sarcophaga talomoensis Magpayo & Kano
Sarcophaga tibialis Macquart
Sarcophaga uliginosa Kramer
Sarcophaga variegata (Scopoli)

Sarcophagula	Sarcophagidae, Diptera
Sarcophagula occidua (Fabricius)	
Sarcoptes	Sarcoptidae, Acari
Sarcoptes bovis auctorum	**see** *Chorioptes bovis*
Sarcoptes cameli Mégnin	**see** *Sarcoptes scabiei*
Sarcoptes canis Gerlach	**see** *Sarcoptes scabiei*
Sarcoptes hominis auctorum	**see** *Sarcoptes scabiei*
Sarcoptes ovis Mégnin	**see** *Sarcoptes scabiei*
Sarcoptes rupicaprae Hering	**see** *Sarcoptes scabiei*
Sarcoptes scabiei (Linnaeus)	
Sarcoptes scabiei bovis auctorum	**see** *Sarcoptes scabiei*
Sarcoptes scabiei cameli Mégnin	**see** *Sarcoptes scabiei*
Sarcoptes scabiei canis Gerlach	**see** *Sarcoptes scabiei*
Sarcoptes scabiei caprae Fürstenberg	**see** *Sarcoptes scabiei*
Sarcoptes scabiei hominis auctorum	**see** *Sarcoptes scabiei*
Sarcoptes scabiei hydrochoeri Mégnin	**see** *Sarcoptes scabiei*
Sarcoptes scabiei ovis Mégnin	**see** *Sarcoptes scabiei*
Sarcoptes scabiei suis Gerlach	**see** *Sarcoptes scabiei*
Sarcoptes scabiei vulpes Fürstenberg	**see** *Sarcoptes scabiei*
Sarcoptes suis Gerlach	**see** *Sarcoptes scabiei*
Sarcosolomonia	Sarcophagidae, Diptera
Sarcosolomonia nathani Lopes & Kano	
Sarcotaces	Philichthyidae, Copepoda
Sarothromyia	Sarcophagidae, Diptera
Sarothromyia indivisa de Souza Lopes	
Sasacarus	Leeuwenhoekiidae, Diptera
Sasacarus panamense Goff	
Sasatrombicula	Trombiculidae, Diptera
Sasatrombicula chejudoensis Goff	
Sathrax	Polyplacidae, Phthiraptera
Sathrax durus Johnson	
Scaphixodes	**see** *Ixodes*
Scaphixodes rothschildi (Nuttall & Warburton)	**see** *Ixodes rothschildi*
Scaptia	Tabanidae, Diptera
Scaptia atra (Philippi)	
Scaptia dorsoguttata (Macquart)	
Scaptia subcontigua (Ferguson)	
Scaptocosa	Lycosidae, Araneae
Scaptocosa raptoria (Walckenaer)	
Scarabaeus	Scarabaeidae, Coleoptera
Scarabaeus carinatus (Gebler)	
Scarabaeus devotus (Redtenbacher)	**see** *Kheper devotus*
Scarabaeus lamarcki Macleay	**see** *Kheper lamarcki*
Scarabaeus rugosus (Hausmann)	
Scarabaeus sacer Linnaeus	
Scarabaeus semipunctatus Fabricius	
Scarabaspis	Eviphididae, Acari
Scarabaspis inexpectatus (Oudemans)	
Scathophaga	Scathophagidae, Diptera
Scathophaga nigripalpis Becker	
Scathophaga scybalaria (Linnaeus)	

Scathophaga stercoraria (Linnaeus)
Scathophaga suilla (Fabricius)
Scathophaga tropicalis Malloch

Sceliphron   Sphecidae, Hymenoptera
Sceliphron caementarium (Drury)
Sceliphron madraspatanum Fabricius
Sceliphron spirifex Linnaeus

Scheloribates   Scheloribatidae, Acari
Scheloribates chauhani Baker
Scheloribates laevigatus (Koch)
Scheloribates latipes (Koch)
Scheloribates pallidulus (Koch)
Schizocarpus   Chirodiscidae, Acari
Schizocarpus mingaudi Trouessart

Schizophthirus   Hoplopleuridae, Phthiraptera
Schizophthirus sicistae Blagoveshchenskiĭ
Schizophthirus singularis Sosnina

Schizopodalges   Psoroptidae, Acari

Schoenbaueria   see Simulium
Schoenbaueria subpusilla (Rubtsov)   see Simulium subpusillum

Schoengastia   Trombiculidae, Acari
Schoengastia baguoiensis Brown & Goff
Schoengastia crossi Nadchatram & Wooster
Schoengastia erinacei Kolebinova
Schoengastia longdongensis Mo et al.
Schoengastia mozambica Kolebinova
Schoengastia sulawesiensis Goff

Schoengastiella   Trombiculidae, Acari
Schoengastiella ligula Radford
Schoengastiella paraconfuciana Wang & Gu
Schoengastiella teras Kolebinova

Schoutedenichia   Trombiculidae, Acari
Schoutedenichia empusa Domrow
Schoutedenichia infrequens Abo-Taka
Schoutedenichia mimema Domrow

Scianophilus   Caligidae, Copepoda

Sciara   Sciaridae, Diptera
Sciara coprophila Lintner   see Bradysia coprophila

Sciomyza   Sciomyzidae, Diptera
Sciomyza dryomyzina Zetterstedt
Sciomyza simplex Fallén
Sciomyza testacea Macquart

Scolopendra   Scolopendridae, Chilopoda
Scolopendra heros Girard
Scolopendra subspinipes dehaanii Brandt
Scolopendra subspinipes subspinipes Leach

Scolopsyllus   Rhopalopsyllidae, Siphonaptera

Scopula   Geometridae, Lepidoptera
Scopula haematophaga Bänziger & Fletcher
Scopula lacriphaga Bänziger & Fletcher

Scopula malayana Bänziger & Fletcher

Scotophaeus
Scotophaeus loricatus (L. Koch)
Gnaphosidae, Araneae

Scutomegninia
Scutomegninia pygmaea Mironov
Avenzoariidae, Acari

Scylla
Scylla serrata (Forskål)
Portunidae, Decapoda

Scytodes
Scytodes cedri Purcell
Scytodidae, Araneae

Sebekia
Sebekia mississippiensis Overstreet *et al.*
Sebekiidae, Pentastomida

Segestria
Segestria florentina (Rossi)
Segestriidae, Araneae

Semenoviana
Semenoviana tamerlana (Saussure)
see *Platycleis*
see *Platycleis tamerlana*

Semiadalia
see *Hippodamia*

Semiothisa
Semiothisa eleonora Stoll
Semiothisa nora Walker
Geometridae, Lepidoptera

Sepedon
Sepedon aenescens Wiedemann
Sepedon sphegea (Fabricius)
Sciomyzidae, Diptera

Sepsis
Sepsis brunnipes (Melander & Spuler)
Sepsis fulgens Hoffmannsegg
Sepsis monostigma Thomson
Sepsis neocynipsea Melander & Spuler
Sepsis punctum (Fabricius)
Sepsis thoracica (Robineau-Desvoidy)
Sepsis violacea Meigen
Sepsidae, Diptera

Sergentomyia
Sergentomyia adleri (Theodor)
Sergentomyia affinis affinis (Theodor)
Sergentomyia affinis vorax (Parrot)
Sergentomyia africana africana (Newstead)
Sergentomyia africana magna (Sinton)
Sergentomyia antennata antennata (Newstead)
Sergentomyia antennata form *sintoni* Pringle
Sergentomyia arpaklensis (Perfil'ev)
Sergentomyia ashfordi Davidson
Sergentomyia babu (Annandale)
Sergentomyia baghdadis (Adler & Theodor)
Sergentomyia bailyi (Sinton)
Sergentomyia bedfordi (Newstead)
Sergentomyia bedfordi media (Kirk & Lewis)
Sergentomyia berentiensis Léger & Rodhain
Sergentomyia buxtoni Theodor
Sergentomyia calcarata Parrot
Sergentomyia caliginosa Davidson
Sergentomyia campester (Sinton)
Sergentomyia christophersi (Sinton)
Sergentomyia clydei (Sinton)
Psychodidae, Diptera

see *Sergentomyia bedfordi*

see *Sergentomyia bailyi*

Sergentomyia congolensis (Bequaert & Walravens)   **see** *Sergentomyia bedfordi*
Sergentomyia dentata (Sinton)
Sergentomyia dhandai Lewis
Sergentomyia dreyfussi (Parrot)   **see** *Grassomyia dreyfussi*
Sergentomyia dubia (Parrot, Mornet & Cadenat)
Sergentomyia fallax (Parrot)
Sergentomyia firmatus (Parrot & Malbrant)   **see** *Sergentomyia bedfordi*
Sergentomyia formica Davidson
Sergentomyia garnhami (Heisch, Guggisberg & Teesdale)
Sergentomyia gobica Artem'ev
Sergentomyia graingeri (Heisch, Guggisberg & Teesdale)
Sergentomyia grekovi (Khodukin)
Sergentomyia hamoni (Abonnenc)
Sergentomyia hunanensis Leng *et al.*
Sergentomyia impudica (Abonnenc)
Sergentomyia indica (Theodor)
Sergentomyia ingrami (Newstead)
Sergentomyia kauli Lewis
Sergentomyia khawi (Raynal)
Sergentomyia kirki (Parrot)
Sergentomyia lushanensis Leng *et al.*
Sergentomyia magna (Sinton)   **see** *Sergentomyia africana magna*
Sergentomyia magnidentata Davidson
Sergentomyia minuta (Rondani)
Sergentomyia minuta sinkiangensis Ting & Ho   **see** *Sergentomyia sinkiangensis*
Sergentomyia murgabiensis (Perfil'ev)
Sergentomyia palestinensis (Adler & Theodor)
Sergentomyia punjabensis (Sinton)
Sergentomyia quanzhouensis Leng & Zhang
Sergentomyia richardi (Parrot & Wanson)
Sergentomyia rima Davidson
Sergentomyia schwetzi (Adler, Theodor & Parrot)
Sergentomyia shorttii (Adler & Theodor)
Sergentomyia silvatica (Raynal & Gaschen)
Sergentomyia sinkiangensis Ting & Ho
Sergentomyia sintoni Pringle   **see** *Sergentomyia antennata* form *sintoni*
Sergentomyia sirohi Kaul, Dhanda & Modi
Sergentomyia sogdiana (Parrot)
Sergentomyia squamipleuris (Newstead)   **see** *Grassomyia squamipleuris*
Sergentomyia squamirostris (Newstead)
Sergentomyia sumbarica (Perfil'ev)
Sergentomyia taizi Lewis
Sergentomyia teteica Artem'ev
Sergentomyia theodori (Parrot)
Sergentomyia tiberiadis (Adler & Theodor)
Sergentomyia turfanensis Hsiung, Guan & Jin
Sergentomyia wuyishanensis Leng & Zhang
Sergentomyia zeylanica (Annandale)
Sergentomyia zumpti (Abonnenc)

Serinirmus   Philopteridae, Phthiraptera
Serinirmus sexytanum Soler Cruz *et al.*

Serradigitus   Vaejovidae, Scorpiones
Serradigitus wupatkiensis (Stahnke)

Serratacarus   Trombiculidae, Acari
Serratacarus dietzi Goff & Whitaker
Serratacarus lasiurus Goff & Whitaker

Serromyia   Ceratopogonidae, Diptera
Serromyia mangrovi Delecolle & Braverman

Sertitympanum *Sertitympanum contiguum* Elsen & Whitaker *Sertitympanum exarmatum* Elsen & Whitaker *Sertitympanum separationis* Elsen & Whitaker	Ameroseiidae, Acari
Shawella *Shawella couloniana* (Saussure)	Blattellidae, Dictyoptera
Sicarius *Sicarius albospinosus* Purcell *Sicarius spatulatus* Pocock	Sicariidae, Araneae
Sigara *Sigara falleni* (Fieber) *Sigara striata* (Linnaeus)	Corixidae, Hemiptera
Sigmactenus *Sigmactenus celebensis* Lewis & Jones	Leptopsyllidae, Siphonaptera
Silpha *Silpha aenescens* Casey *Silpha carinata* Herbst *Silpha noveboracensis* Forster *Silpha obscura* Linnaeus	Silphidae, Coleoptera **see** *Heterosilpha aenescens* **see** *Oiceoptoma noveboracense*
Silvius *Silvius latifrons* Olsuf'ev *Silvius oshimaensis* Hayakawa, Takahasi & Suzuki *Silvius shirakii* Philip & Mackerras	Tabanidae, Diptera
Simocephalus *Simocephalus vetulus* (O.F. Müller)	Daphniidae, Branchiopoda
Simploce *Simploce capitata* (Saussure) *Simploce hospes* Perkins *Simploce pallens* (Stephens)	**see** *Symploce* **see** *Symploce capitata* **see** *Symploce capitata* **see** *Symploce pallens*
Simulium *Simulium adersi* Pomeroy *Simulium admixtum* Craig *Simulium aequifurcatum* Lutz *Simulium aestivum* Davies, Peterson & Wood *Simulium alajense* Rubtsov *Simulium albellum* Rubtsov *Simulium albivirgulatum* Wanson & Henrard *Simulium almae* (Yankovskiĭ & Koshkimbaev) *Simulium amazonicum* Goeldi *Simulium anatolicum* Craig *Simulium angustipes* Edwards *Simulium angustitarse angustitarse* (Lundström) *Simulium angustitarse zaporojae* (Pavlichenko) *Simulium antillarum* Jennings *Simulium aokii* (Takahasi) *Simulium arakawae* Matsumura *Simulium arcticum* Malloch *Simulium argenteostriatum* Strobl *Simulium argentiscutum* Shelley & Dias *Simulium argus* Williston *Simulium argyreatum* Meigen *Simulium arlecchinum* Craig *Simulium aureohirtum* Brunetti *Simulium aureum* Fries *Simulium auricoma* Meigen	Simuliidae, Diptera

Simulium azorense Carlsson
Simulium barraudi Puri
Simulium bezzii (Corti)
Simulium bicorne Dorogostaĭskiĭ, Rubtsov & Vlasenko
Simulium bidentatum (Shiraki)
Simulium bivittatum Malloch
Simulium bonaerense Coscarón & Wygodzinski
Simulium bovis De Meillon
Simulium brachycladum Lutz & Pinto
Simulium bravermani Beaucournu–Saguez
Simulium bulla Davies & Györkös
Simulium buxtoni Austen
Simulium callidum (Dyar & Shannon)
Simulium canadense Hearle
Simulium canonicolum (Dyar & Shannon)
Simulium carpathicum (Knoz)
Simulium cataractarum Craig
Simulium cauchense Floch & Abonnenc
Simulium cervicornutum Pomeroy
Simulium ceylonicum Enderlein
Simulium chaquense Coscarón
Simulium cheesmanae Edwards
Simulium chiriquiense Field
Simulium cholodkovskii Rubtsov
Simulium chutteri Lewis
Simulium condici (Baranov)
Simulium congareenarum (Dyar & Shannon)
Simulium corbis Twinn
Simulium costatum Friederichs
Simulium covagarciai Ramírez Pérez *et al.*
Simulium cremnosi Davies & Györkös
Simulium cryophilum (Rubtsov)
Simulium cuasisanguineum Ramírez Pérez *et al.*   **see** *Simulium oyapockense*
Simulium cuneatum (Enderlein)
Simulium daisense (Takahasi)
Simulium damascenoi Py–Daniel
Simulium damnosum Theobald
Simulium decorum Walker
Simulium defoliarti Stone & Peterson
Simulium delponteianum Wygodzinsky
Simulium dentatum Puri
Simulium dentulosum altissimum Fain, Bafort & Silberstein
Simulium dentulosum dentulosum Roubaud
Simulium dentulosum trifurcatum Fain, Bafort & Silberstein
Simulium dieguerense Vajime & Dunbar
Simulium diversifurcatum Lutz   **see** *Simulium subnigrum*
Simulium dola Davies & Györkös
Simulium dorieri Doby & Rault   **see** *Simulium monticola*
Simulium duboisi Fain
Simulium ela Davies & Györkös
Simulium emarginatum Davies, Peterson & Wood
Simulium ephemerophilum Rubtsov
Simulium equinum (Linnaeus)
Simulium erimoense (Ono)
Simulium erythrocephalum (DeGeer)
Simulium euryadminiculum Davies
Simulium exiguum Roubaud
Simulium fallisi (Golini)
Simulium fibrinflatum Twinn
Simulium fulvinotum Cerqueira & Nunes de Melo
Simulium galeratum Edwards   **see** *Simulium reptans*
Simulium galloprovinciale Guidicelli
Simulium gariepense De Meillon
Simulium ghoomense Datta

Simulium goeldii Cerqueira & Nunes de Mello
Simulium gonzalezi Vargas & Díaz Nájera
Simulium gracilis Datta
Simulium griseicolle Becker
Simulium grisescens Brunetti
Simulium guianense Wise
Simulium guimari Becker
Simulium haematopotum Malloch
Simulium hargreavesi Gibbins
Simulium hieroglyphicum Peterson *et al.*
Simulium himalayense Puri
Simulium hirsutum Pomeroy
Simulium horacioi Okazawa & Onishi
Simulium horokaense Ono
Simulium hunteri Malloch
Simulium ibericum Crosskey & Santos Grácio
Simulium ignescens Roubaud
Simulium inaequale Peterson & Shannon
Simulium incrustatum Lutz
Simulium intermedium Roubaud
Simulium irakae Smart   **see** *Simulium buxtoni*
Simulium iwatense (Shiraki)
Simulium japonicum Matsumura
Simulium jenningsi Malloch
Simulium juxtadamnosum Gouteux
Simulium katmai Dyar & Shannon   **see** *Simulium decorum*
Simulium kawamurae Matsumura
Simulium keiseri (Rubtsov)
Simulium kilibanum Gouteux
Simulium konoi (Takahasi)
Simulium krombeini Davies & Györkös
Simulium languidum Davies & Györkös
Simulium larvipilosum Okazawa
Simulium latipes (Meigen)
Simulium latizonum (Rubtsov)
Simulium limay Wygodzinsky
Simulium limbatum Knab
Simulium lineatum (Meigen)
Simulium loerchae Adler
Simulium longipalpe Bel'tyukova
Simulium lourencoi Py-Daniel
Simulium luggeri Nicholson & Mickel
Simulium machadoi Luna de Carvalho
Simulium machadoi Ramírez Pérez   **see** *Simulium aequifurcatum*
Simulium mbarigui Coscarón & Wygodzinsky   **see** *Simulium subnigrum*
Simulium mcmahoni De Meillon
Simulium mediterraneum Puri   **see** *Simulium pseudequinum*
Simulium meridionale Riley
Simulium metallicum Bellardi
Simulium mexicanum Bellardi
Simulium mie Ogata & Sasa
Simulium minusculum Lutz
Simulium monticola Friederichs
Simulium morsitans Edwards
Simulium morsitans longipalpe Bel'tyukova   **see** *Simulium longipalpe*
Simulium neavei Roubaud
Simulium neornatipes Dumbleton
Simulium nigrimanum Macquart
Simulium nigritarse Coquillett
Simulium niha Giudicelli & Dia
Simulium nikkoense Shiraki
Simulium nishijimai (Ono)
Simulium nitidifrons Edwards
Simulium nitidithorax Puri

Simulium nodosum Puri
Simulium noelleri Friederichs
Simulium nubis Davies & Györkös
Simulium obikumbense (Rubtsov)   see Simulium ephemerophilum
Simulium ochraceum Walker
Simulium ogatai (Rubtsov)
Simulium oresti (Vorobets)
Simulium ornatum Meigen
Simulium oshimaense (Ono)
Simulium oviceps Edwards
Simulium ovtshinnikovi Rubtsov
Simulium oyapockense Floch & Abonnenc
Simulium panamense Fairchild
Simulium paraloutetense Crosskey
Simulium pattoni Senior White
Simulium paynei Vargas
Simulium perflavum Roubaud
Simulium pertinax Kollar
Simulium petricolum (Rivosecchi)
Simulium pictipes Hagen
Simulium pinhaoi Grácio
Simulium pintoi d'Andretta & d'Andretta
Simulium piperi Dyar & Shannon
Simulium pontinum Rivosecchi
Simulium posticatum Meigen
Simulium praelargum Datta
Simulium pruinosum Lutz   see Simulium nigrimanum
Simulium pseudequinum Séguy
Simulium pseudoamazonicum Ramírez Pérez & Peterson
Simulium pseudosanguineum Ramírez Pérez & Peterson
Simulium pulverulentum Knab
Simulium purii Datta
Simulium quadrivittatum Loew
Simulium quinquestriatum (Shiraki)
Simulium ramosum Puri
Simulium rangeli Ramírez-Pérez et al.   see Simulium cauchense
Simulium rashidi Lewis
Simulium rasyani Garms, Kerner & Meredith
Simulium rendalense (Golini)
Simulium reptans galeratum Edwards   see Simulium reptans
Simulium reptans reptans (Linnaeus)
Simulium rithrogenophilum (Konurbaev)
Simulium roraimense Nunes de Mello
Simulium rostratum (Lundström)
Simulium rotundum Gibbins
Simulium rubicundulum Knab
Simulium rubiginosum (Enderlein)
Simulium rubrithorax Lutz
Simulium rufibasis Brunetti
Simulium ruficorne Macquart
Simulium salopiense Edwards   see Simulium lineatum
Simulium sanchezi Ramírez Pérez et al.   see Simulium oyapockense
Simulium sanctipauli Vajime & Dunbar
Simulium sanguineum Knab
Simulium schoutedeni Wanson
Simulium securiforme (Rubtsov)
Simulium shevtshenkovae Rubtsov
Simulium shevyakovi Dorogostaisky et al.
Simulium shogakii (Rubtsov)
Simulium siolii Py-Daniel
Simulium sirbanum Vajime & Dunbar
Simulium slossonae Dyar & Shannon
Simulium soubrense Vajime & Dunbar
Simulium spinosum Doby & Deblock

Simulium squamosum (Enderlein)
Simulium subcostatum (Takahasi)
Simulium sublacustre Davies   **see** Simulium rostratum
Simulium subnigrum Lutz
Simulium subpusillum (Rubtsov)
Simulium sudanense Vajime & Dunbar
Simulium suzukii Rubtsov
Simulium tahitiense Edwards
Simulium takahasii (Rubtsov)
Simulium talassicum (Yankovskiǐ)
Simulium tallaferroae Ramírez Pérez
Simulium tenerificum Crosskey
Simulium teruamanga Craig & Craig
Simulium travassosi d'Andretta & d'Andretta
Simulium tridens Freeman & De Meillon
Simulium trirugosum Davies & Györkös
Simulium tuberosum (Lundström)
Simulium uchidai (Takahasi)
Simulium unicornutum Pomeroy
Simulium usovae (Golini)
Simulium varicorne Edwards
Simulium variegatum Meigen
Simulium veltistshevi Rubtsov
Simulium venezuelense Ramírez Pérez & Peterson
Simulium venustum Say
Simulium verecundum Stone & Jamnback
Simulium vernum Macquart
Simulium vittatum Zetterstedt
Simulium voilensis Serban
Simulium vorax Pomeroy
Simulium vulgare Rubtsov
Simulium wolffhuegeli (Enderlein)
Simulium yahense Vajime & Dunbar
Simulium yarzabali Ramírez-Pérez
Simulium yerburyi Edwards   **see** Simulium latipes
Simulium yonagoense Okamoto
Simulium yonakuniense Takaoka

Sinopotamon   Sinopotamidae, Decapoda
Sinopotamon denticulatum (Milne-Edwards)

Siphona   Tachinidae, Diptera
Siphona exigua (de Meijere)   **see** Haematobia irritans exigua

Siphunculina   Chloropidae, Diptera
Siphunculina minima de Meijere

Sirex   Siricidae, Hymenoptera
Sirex juvencus (Linnaeus)

Sirthenea   Reduviidae, Hemiptera
Sirthenea amazona Stål
Sirthenea carinata (Fabricius)
Sirthenea flavipes (Stål)
Sirthenea stria (Fabricius)

Sisyphus   Scarabaeidae, Coleoptera
Sisyphus bornemisszanus Endrődi
Sisyphus schaefferi (Linnaeus)
Sisyphus spinipes (Thunberg)

Siteroptes   Pygmephoridae, Acari
Siteroptes muscarius Martin

Smittia *Smittia pratorum* (Goetghebuer)	Chironomidae, Diptera
Solenopotes *Solenopotes burmeisteri* (Fahrenholz) *Solenopotes capillatus* Enderlein *Solenopotes capreoli* Freund *Solenopotes ferrisi* (Fahrenholz)	Linognathidae, Phthiraptera
Solenopsis *Solenopsis aurea* Wheeler *Solenopsis geminata geminata* (Fabricius) *Solenopsis geminata rufa* (Jerdon) *Solenopsis invicta* Buren *Solenopsis maniosa* Wheeler *Solenopsis richteri* Forel *Solenopsis saevissima* (F. Smith) *Solenopsis xyloni* McCook	Formicidae, Hymenoptera **see** *Solenopsis xyloni*
Soriculopus *Soriculopus lukoschusi* Haitlinger	Glycyphagidae, Acari
Spalacarus *Spalacarus elongatus* Fain & Lukoschus	Listrophoridae, Acari
Spalangia *Spalangia cameroni* Perkins *Spalangia endius* Walker *Spalangia haematobiae* Ashmead *Spalangia nigra* Latreille *Spalangia nigroaenea* Curtis	Pteromalidae, Hymenoptera
Spastomeloe *Spastomeloe weyrauchi* (Kaszab)	Meloidae, Coleoptera
Spelaeorhynchus *Spelaeorhynchus hutsoni* Martin	Spelaeorhynchidae, Acari
Speleocola *Speleocola tamarina* Goff, Whitaker & Dietz	Trombiculidae, Acari
Speleognathopsis *Speleognathopsis galli* Cooreman	Ereynetidae, Acari
Speleognathus *Speleognathus australis* Womersley	Ereynetidae, Acari
Speleorodens *Speleorodens clethrionomys* (Fain & Lukoschus)	Ereynetidae, Acari
Spelobia *Spelobia bifrons* (Stenhammar) *Spelobia elegans* (Spuler)	Sphaeroceridae, Diptera **see** *Spelobia bifrons*
Speophyes *Speophyes lucidulus* (Delarouzée)	Leiodidae, Coleoptera
Sphaeridium *Sphaeridium lunatum* Fabricius *Sphaeridium scarabaeoides* (Linnaeus)	Hydrophilidae, Coleoptera
Sphaerobothria *Sphaerobothria hoffmanni* Karsch	Theraphosidae, Araneae

Sphaerocera *Sphaerocera curvipes* Latreille *Sphaerocera pseudomonilis* Nishijima & Yamazaki	Sphaeroceridae, Diptera
Sphaerodema *Sphaerodema nepoides* (Fabricius) *Sphaerodema rusticum* (Fabricius)	**see** *Diplonychus* **see** *Diplonychus nepoides* **see** *Diplonychus rusticus*
Sphaeroma *Sphaeroma peruvianum* Richardson	Sphaeromatidae, Isopoda
Sphaeromias	Ceratopogonidae, Diptera
Sphecophaga *Sphecophaga vesparum burra* Cresson *Sphecophaga vesparum vesparum* Curtis	Ichneumonidae, Hymenoptera
Sphyrion *Sphyrion lumpi* (Krøyer)	Sphyriidae, Copepoda
Spilopsyllus *Spilopsyllus cuniculi* (Dale)	Pulicidae, Siphonaptera
Spinturnix *Spinturnix acuminatus* (Koch) *Spinturnix brevisetosus* Gu & Wang *Spinturnix domrowi* Deunff & Volleth *Spinturnix kolenatii* Oudemans *Spinturnix myoti* (Kolenati) *Spinturnix mystacinus* (Kolenati) *Spinturnix plecotinus* (Koch) *Spinturnix semilunaris* De Meillon & Lavoipierre *Spinturnix sinicus* Gu & Wang *Spinturnix vespertilionis*	Spinturnicidae, Acari
Squamulotilla	Mutillidae, Hymenoptera
Staphylinus *Staphylinus olens* Müller	Staphylinidae, Coleoptera
Steatoda *Steatoda bipunctata* (Linnaeus) *Steatoda paykulliana* (Walckenaer)	Theridiidae, Araneae
Steatonyssus *Steatonyssus periblepharus* Kolenati *Steatonyssus viator* (Hirst)	Macronyssidae, Acari
Stegopterna *Stegopterna mutata* (Malloch) *Stegopterna nukabirana* Ono	Simuliidae, Diptera
Stelopolybia *Stelopolybia multipicta* (Haliday) *Stelopolybia pallipes* (Olivier)	Vespidae, Hymenoptera
Stenepteryx *Stenepteryx hirundinis* (Linnaeus)	**see** *Crataerina* **see** *Crataerina hirundinis*
Stenistomera *Stenistomera alpina* (Baker)	Hystrichopsyllidae, Siphonaptera
Stenoponia *Stenoponia tripectinata* (Tiraboschi)	Hystrichopsyllidae, Siphonaptera

Sternolophus *Sternolophus rufipes* (Fabricius)	Hydrophilidae, Coleoptera
Sternopsylla *Sternopsylla distincta distincta* (Rothschild) *Sternopsylla distincta speciosa* Johnson *Sternopsylla distincta texana* (C. Fox)	Ischnopsyllidae, Siphonaptera
Sternostoma *Sternostoma tracheacolum* Lawrence	Rhinonyssidae, Acari
Stigmaeus *Stigmaeus sinai* Swift	Stigmaeidae, Acari
Stilbomyella	Calliphoridae, Diptera
Stilobezzia *Stilobezzia donskoffi* Clastrier	Ceratopogonidae, Diptera
Stivalius *Stivalius exoticus* Lewis & Jones	Pygiopsyllidae, Siphonaptera
Stomorhina *Stomorhina discolor* (Fabricius)	Calliphoridae, Diptera
Stomoxys *Stomoxys calcitrans* (Linnaeus) *Stomoxys niger* Macquart *Stomoxys sitiens* Rondani	Muscidae, Diptera
Stonemyia *Stonemyia enokizonoi enokizonoi* Ôuchi *Stonemyia enokizonoi ishizuchiensis* Yonetsu *Stonemyia yezoensis* (Shiraki)	Tabanidae, Diptera
Stratiomys *Stratiomys badia* Walker	Stratiomyidae, Diptera
Strebla	Streblidae, Diptera
Strelkoviacarus *Strelkoviacarus integer* (Trouessart & Neumann)	Epidermoptidae, Acari
Strobiloestrus	Oestridae, Diptera
Stromatium *Stromatium fulvum* (Villers)	Cerambycidae, Coleoptera
Sturnophagoides *Sturnophagoides brasiliensis* Fain *Sturnophagoides halterophilus* Fain & Feinberg	Pyroglyphidae, Acari
Styloconops *Styloconops hamariensis* Herzi & Sabatini *Styloconops spinosifrons* (Carter)	**see** *Leptoconops* **see** *Leptoconops hamariensis* **see** *Leptoconops spinosifrons*
Stypommisa *Stypommisa anoriensis* Fairchild & Wilkerson *Stypommisa apicalis* Fairchild & Wilkerson *Stypommisa aripuana* Fairchild & Wilkerson *Stypommisa changena* Fairchild & Wilkerson *Stypommisa kroeberi* Fairchild & Wilkerson *Stypommisa spilota* Fairchild & Wilkerson *Stypommisa xanthicornis* Fairchild & Wilkerson	Tabanidae, Diptera

Suidasia	Winterschmidtiidae, Acari
Suidasia medanensis Oudemans	**see** *Suidasia pontifica*
Suidasia nesbitti Hughes	
Suidasia pontifica Oudemans	
Sulcicnephia	**see** *Simulium*
Sulcicnephia ovtshinnikovi (Rubtsov)	**see** *Simulium ovtshinnikovi*
Supella	Blattellidae, Dictyoptera
Supella longipalpa (Fabricius)	
Supella supellectilium (Serville)	**see** *Supella longipalpa*
Susanomira	Sepsidae, Diptera
Susanomira caucasica Pont	
Sylvicola	Anisopodidae, Diptera
Sylvicola cinctus (Fabricius)	
Sylvicola fenestralis (Scopoli)	
Symphoromyia	Rhagionidae, Diptera
Symphoromyia crassicornis (Panzer)	
Symphoromyia hirta Johnson	
Symphoromyia spitzeri Chvála	
Symploce	Blattellidae, Dictyoptera
Symploce capitata (Saussure)	
Symploce hospes (Perkins)	**see** *Symploce capitata*
Symploce pallens (Stephens)	
Synapsis	Scarabaeidae, Coleoptera
Synapsis cambeforti Krikken	
Synapsis ritsemae Lansberge	
Synapsis thoas Sharp	
Synopsyllus	Pulicidae, Siphonaptera
Synopsyllus fonquerniei Wagner & Roubaud	
Synosternus	Pulicidae, Siphonaptera
Synosternus pallidus (Taschenberg)	
Synthesiomyia	Muscidae, Diptera
Synthesiomyia nudiseta (Wulp)	
Syntomosphyrum	Eulophidae, Hymenoptera
Syntomosphyrum glossinae Waterston	**see** *Nesolynx glossinae*
Syringanoetus	Histiostomatidae, Acari
Syringanoetus faini Mahunka & Eraky	
Syringophilus	Syringophilidae, Acari
Syringophilus bipectinatus Heller	
Syritta	Syrphidae, Diptera
Syritta pipiens (Linnaeus)	
Syscenus	Aegidae, Isopoda
Syscenus atlanticus Kononenko	

T

Tabanus	Tabanidae, Diptera
Tabanus abactor Philip	
Tabanus agrestis Wiedemann	**see** *Atylotus agrestis*
Tabanus albifrons Szilady	
Tabanus amaenus Walker	

Tabanus americanus Förster
Tabanus appendicifer Szilady
Tabanus appendiculatus Szilady   **see** *Tabanus appendicifer*
Tabanus armeniacus (Kröber)
Tabanus atamuradovi Dolin & Andreeva
Tabanus atratoides Burger
Tabanus atratus Fabricius
Tabanus atropathenicus Olsuf'ev
Tabanus atropilosus Burger
Tabanus autumnalis autumnalis Linnaeus
Tabanus autumnalis brunnescens Szilady   **see** *Tabanus autumnalis*
Tabanus besti Surcouf
Tabanus bifarius Loew
Tabanus biguttatus Wiedemann
Tabanus bovinus Linnaeus
Tabanus bromius Linnaeus
Tabanus calens Linnaeus
Tabanus calidus Walker
Tabanus camelarius Austen
Tabanus catenatus Walker
Tabanus caucasius Kröber
Tabanus chrysurinus (Enderlein)
Tabanus chrysurus Loew
Tabanus cognatus Loew   **see** *Tabanus glaucopis*
Tabanus colchidicus Olsuf'ev
Tabanus colon Thunberg
Tabanus conformis Walker
Tabanus conterminus Walker
Tabanus cordiger Meigen
Tabanus crassus Walker
Tabanus cymatophorus Osten Sacken
Tabanus dolini Ivanishchuk
Tabanus dorsiger dorsiger Wiedemann
Tabanus dorsiger dorsovittatus Macquart   **see** *Tabanus occidentalis dorsovittatus*
Tabanus dorsiger stenocephalus Hine
Tabanus dorsilinea Wiedemann
Tabanus equalis Hine
Tabanus fraseri Austen
Tabanus fulvicallus Philip
Tabanus fulvimedioides Shiraki
Tabanus fuscicostatus Hine
Tabanus gladiator Stone
Tabanus glaucopis glaucopis Meigen
Tabanus glaucopis rubra (Mushamp)
Tabanus gratus Loew
Tabanus hauseri Olsuf'ev
Tabanus hybridus Wiedemann
Tabanus importunus Wiedemann
Tabanus indrae Hauser
Tabanus infestus Bogachev & Samedov
Tabanus iyoensis Shiraki   **see** *Hirosia iyoensis*
Tabanus karaosus Timmer
Tabanus katoi Kôno & Takahasi
Tabanus kinoshitai Kôno & Takahasi
Tabanus laetetinctus Becker
Tabanus laticornis Hine
Tabanus leleani Austen
Tabanus leucostomus Loew
Tabanus lineola hinellus Philip
Tabanus lineola lineola Fabricius
Tabanus lubutuensis Bequaert
Tabanus macdonaldi Philip
Tabanus macer (Bigot)   **see** *Tabanus dorsilinea*
Tabanus maculicornis Zetterstedt

Tabanus mandarinus Schiner
Tabanus mateusi Dias
Tabanus medionotatus Austen
Tabanus miki colchidicus Olsuf'ev   see *Tabanus colchidicus*
Tabanus miki miki Brauer
Tabanus mordax Austen
Tabanus mossambicensis Dias
Tabanus mularis Stone
Tabanus nigrovittatus Macquart
Tabanus nipponicus Murdoch & Takahasi
Tabanus nyasae Ricardo
Tabanus obsolescens Pandelle
Tabanus occidentalis dorsovittatus Macquart
Tabanus occidentalis occidentalis Linnaeus
Tabanus pallidiventris Olsuf'ev   see *Tabanus amaenus*
Tabanus par Walker
Tabanus persimilis Dolin & Andreeva
Tabanus punctifer Osten Sacken
Tabanus quatuornotatus Meigen
Tabanus quinquevittatus Wiedemann
Tabanus regularis Jaennicke
Tabanus riyadhae Amoudi & Leclercq
Tabanus rubidus Wiedemann
Tabanus rufidens (Bigot)
Tabanus rupium Brauer
Tabanus sackeni Fairchild
Tabanus samawangensis Burger
Tabanus sapporoensis Shiraki   see *Hirosia sapporoensis*
Tabanus secedens Walker
Tabanus selvaticus Burger *et al.*
Tabanus semenovi Olsuf'ev
Tabanus shelkovnikovi Paramonov
Tabanus sierrensis Burger *et al.*
Tabanus similis Macquart
Tabanus simpsoni Austen
Tabanus simulans Walker   see *Tabanus nigrovittatus*
Tabanus spectabilis Loew
Tabanus spodopterus Meigen
Tabanus striatus Fabricius
Tabanus subhybridus Philip
Tabanus subsimilis Bellardi
Tabanus sudeticus Zeller
Tabanus sufis Jaennicke
Tabanus sulcifrons Macquart
Tabanus taeniola Palisot de Beauvois
Tabanus taiwanus Hayakawa & Takahasi
Tabanus tergestinus Eggers
Tabanus thoracinus Palisot de Beauvois
Tabanus tokaraensis Hayakawa & Suzuki
Tabanus tokunoshimaensis Hayakawa & Suzuki
Tabanus trigeminus Coquillett
Tabanus trigonus Coquillett
Tabanus trimaculatus Palisot de Beauvois
Tabanus triquetrornatus Carter
Tabanus vicarius Walker   see *Tabanus quinquevittatus*
Tabanus vittiger Thomson
Tabanus vivax Osten Sacken
Tabanus yanbaruensis Hayakawa & Yoneyama
Tabanus yoneyamai Hayakawa
Tabanus zimini Olsuf'ev

Tachinaephagus   Encyrtidae, Hymenoptera
Tachinaephagus stomoxicida Subba Rao
Tachinaephagus zealandicus Ashmead

Tainanina	Calliphoridae, Diptera
Tanypus	Chironomidae, Diptera
Tanypus grodhausi Sublette	**see** *Tanypus nubifer*
Tanypus nubifer Skuse	
Tanypus punctipennis Meigen	
Tanytarsus	Chironomidae, Diptera
Tanytarsus bathophilus Kieffer	
Tanytarsus dissimilis Johannsen	**see** *Paratanytarsus grimmii*
Tanytarsus gracilentus Holmgren	
Tanytarsus oyamai Sasa	
Tanytarsus sylvaticus (van der Wulp)	
Tapinoma	Formicidae, Hymenoptera
Tapinoma melanocephalum (Fabricius)	
Tarentula	Lycosidae, Araneae
Tarentula cubensis Farrington	
Tarnetrum	Libellulidae, Odonata
Tarnetrum corruptum (Hagen)	
Tarsonemoides	Tarsonemidae, Acari
Tarsonemoides noxius (Humiczewska)	**see** *Tarsonemus noxius*
Tarsonemus	Tarsonemidae, Acari
Tarsonemus granarius Lindquist	
Tarsonemus minusculus (Canestrini & Fanzago)	
Tarsonemus noxius (Humiczewska)	
Tarsonemus rakowiensis (Kropczynska)	
Tarsopsylla	Ceratophyllidae, Siphonaptera
Tarsopsylla octodecimdentata coloradensis (Baker)	
Tarsopsylla octodecimdentata octodecimdentata (Kolenati)	
Tateracarus	Trombiculidae, Acari
Tateracarus quadrisetosus Goff	
Taurocerastes	Geotrupidae, Coleoptera
Taurocerastes patagonicus Philippi	
Tectocepheus	Tectocepheidae, Acari
Tectocepheus sarekensis (Trägårdh)	
Tectocepheus velatus (Michael)	
Tegenaria	Agelenidae, Araneae
Tegenaria agrestis (Walckenaer)	
Tegenaria atrica Koch	
Tegenaria gigantea Chamberlin & Ivie	
Teichomyza	**see** *Scatella*
Teichomyza fusca Macquart	**see** *Scatella fusca*
Teinocoptes	Teinocoptidae, Acari
Teinocoptes amphipterinon Klompen *et al.*	
Telenomus	Scelionidae, Hymenoptera
Telenomus angustatus (Thomson)	
Telenomus capito De Santis & Loiácono	
Telenomus costalimai Ortiz & Alvarez	
Telenomus dimmocki Ashmead	**see** *Telenomus podisi*
Telenomus inclinis Le	
Telenomus podisi Ashmead	

Telenomus zeli Johnson

Teleogryllus　　　　　　　　　　　　　　　　Gryllidae, Orthoptera
Teleogryllus oceanicus (Le Guillou)

Telmatoscopus　　　　　　　　　　　　　　　Psychodidae, Diptera
Telmatoscopus albipunctatus (Williston)

Telomerina　　　　　　　　　　　　　　　　Sphaeroceridae, Diptera
Telomerina beringiensis Marshall

Tentyria　　　　　　　　　　　　　　　　　Tenebrionidae, Coleoptera
Tentyria gigas Faldermann

Tephrochlamys　　　　　　　　　　　　　　Heleomyzidae, Diptera
Tephrochlamys japonica Okadome

Terrilimosina　　　　　　　　　　　　　　Sphaeroceridae, Diptera
Terrilimosina deemingi Marshall
Terrilimosina smetanai Marshall
Terrilimosina unio Marshall

Tetanocera　　　　　　　　　　　　　　　　Sciomyzidae, Diptera
Tetanocera ferruginea Fallén

Tetisimulium　　　　　　　　　　　　　　**see** *Simulium*
Tetisimulium alajense (Rubtsov)　　　　**see** *Simulium alajense*
Tetisimulium bezzii (Corti)　　　　　　**see** *Simulium bezzii*
Tetisimulium condici (Baranov)　　　　**see** *Simulium condici*

Tetramorium　　　　　　　　　　　　　　　Formicidae, Hymenoptera
Tetramorium caespitum (Linnaeus)
Tetramorium insolens F. Smith
Tetramorium lanuginosum (Mayr)
Tetramorium simillimum (F. Smith)

Tetraponera　　　　　　　　　　　　　　　Formicidae, Hymenoptera
Tetraponera rufonigra (Jerdon)

Tetrapsyllus　　　　　　　　　　　　　　　Rhopalopsyllidae, Siphonaptera
Tetrapsyllus maulinus Beaucournu & Gallardo N.
Tetrapsyllus rhombus Smit
Tetrapsyllus tantillus Jordan & Rothschild

Tetrastichus　　　　　　　　　　　　　　　Eulophidae, Hymenoptera
Tetrastichus asthenogmus (Waterston)　**see** *Aprostocetus asthenogmus*
Tetrastichus dytisciarum Kostyukov & Fursov
Tetrastichus hagenowii (Ratzeburg)　　**see** *Aprostocetus hagenowii*
Tetrastichus natans Kostyukov & Fursov

Thaumetopoea　　　　　　　　　　　　　　Notodontidae, Lepidoptera
Thaumetopoea bonjeani Powell
Thaumetopoea pityocampa (Denis & Schiffermüller)
Thaumetopoea processionea (Linnaeus)
Thaumetopoea solitaria (Freyer)
Thaumetopoea wilkinsoni Tams

Thelphusa　　　　　　　　　　　　　　　　Sinopotamidae, Decapoda
Thelphusa denticulatum Milne-Edwards　**see** *Sinopotamon denticulatum*

Thelyphassa　　　　　　　　　　　　　　　Oedemeridae, Coleoptera
Thelyphassa lineata (Fabricius)

Theobaldia *Theobaldia longiareolata* (Macquart)	**see** *Culiseta* **see** *Culiseta longiareolata*
Thereva *Thereva annulata* Fabricius	Therevidae, Diptera
Theridion *Theridion blackwalli* O. Pickard–Cambridge *Theridion contreras* Levi *Theridion rufipes* Lucas	Theridiidae, Araneae **see** *Nesticodes rufipes*
Theridiosoma *Theridiosoma globosum* (O. Pickard–Cambridge)	Theridiosomatidae, Araneae **see** *Epilineutes globosus*
Therioplectes	Tabanidae, Diptera
Thermobia *Thermobia domestica* (Packard)	Lepismatidae, Thysanura
Thermocyclops *Thermocyclops decipiens* Kiefer *Thermocyclops dybowskii* (Landé) *Thermocyclops emini* (Mrázek) *Thermocyclops incisus* Kiefer *Thermocyclops infrequens infrequens* (Kiefer) *Thermocyclops infrequens nigerianus* (Kiefer) *Thermocyclops inopinus* (Kiefer) *Thermocyclops nigerianus* (Kiefer) *Thermocyclops oblongatus nigerianus* (Kiefer) *Thermocyclops oblongatus oblongatus* *Thermocyclops oithonoides* (Sars)	Cyclopidae, Copepoda **see** *Thermocyclops oblongatus oblongatus* **see** *Thermocyclops oblongatus nigerianus* **see** *Thermocyclops oblongatus nigerianus*
Thermonetus *Thermonetus basillaris* (Harris)	Dytiscidae, Coleoptera
Thersitina *Thersitina gasterostei* (Pagenstecher)	Ergasilidae, Copepoda
Thliptoceras *Thliptoceras anthropophilum* Bänziger *Thliptoceras lacriphagum* Bänziger *Thliptoceras umoremsugente* Bänziger	Pyralidae, Lepidoptera
Thomazomyia *Thomazomyia adunca* Souza Lopes *Thomazomyia fluminensis* Souza Lopes	Sarcophagidae, Diptera
Thomomydoecus *Thomomydoecus byersi* Hellenthal & Price	Trichodectidae, Phthiraptera
Thoracochaeta *Thoracochaeta johnsoni* (Spuler)	Sphaeroceridae, Diptera
Thorectes *Thorectes albarracinus* Wagner *Thorectes intermedius* (Costa) *Thorectes laevigatus cobosi* (Baraud) *Thorectes laevigatus laevigatus* (Fabricius) *Thorectes sericeus* Jekel	**see** *Geotrupes* **see** *Geotrupes albarracinus* **see** *Geotrupes intermedius* **see** *Geotrupes laevigatus cobosi* **see** *Geotrupes laevigatus laevigatus* **see** *Geotrupes sericeus*
Thyreophora *Thyreophora anthropophaga* Robineau–Desvoidy	Piophilidae, Diptera **see** *Centrophlebomyia anthropophaga*
Thyromyobia	Myobiidae, Acari

Thysanocercus	Alloptidae, Acari
Thysanocercus affinis Gaud & Peterson	
Thysanocercus brachyurus Gaud & Peterson	
Thysanocercus callithyrus Gaud & Mouchet	**see** *Thysanocercus cypseli*
Thysanocercus cerionotus Gaud & Peterson	
Thysanocercus cypseli (Canestrini & Berlese)	
Thysanocercus parvulus Gaud & Peterson	
Thysanocercus tanycnemus Gaud & Peterson	
Tiamastus	Rhopalopsyllidae, Siphonaptera
Tiamastus callens (Jordan & Rothschild)	
Tidarren	Theridiidae, Araneae
Tidarren sisyphoides (Walckenaer)	
Tigriopus	Harpacticidae, Copepoda
Tigriopus brevicornis Müller	
Tinaminyssus	Dermanyssidae, Acari
Tinaminyssus melloi (Castro)	
Tinea	Tineidae, Lepidoptera
Tinea pellionella (Linnaeus)	
Tineola	Tineidae, Lepidoptera
Tineola bisselliella (Hummel)	
Tiniocellus	Scarabaeidae, Coleoptera
Tiniocellus panthera Boucomont	
Tiniocellus spinipes (Roth)	
Tipula	Tipulidae, Diptera
Tipula abdominalis (Say)	
Tityopsis	Buthidae, Scorpiones
Tityus	Buthidae, Scorpiones
Tityus asthenes Pocock	
Tityus atriventer Pocock	
Tityus bahiensis (Perty)	
Tityus bastosi Lourenço	
Tityus cambridgei Pocock	
Tityus clathratus Koch	
Tityus forcipula (Gervais)	
Tityus insignis (Pocock)	
Tityus jussarae Lourenço	
Tityus neglectus Mello-Leitão	
Tityus pictus Pocock	
Tityus pugilator Pocock	
Tityus serrulatus Lutz & Mello	
Tityus silvestris Pocock	
Tityus stigmurus (Thorell)	
Tityus trinitatis Pocock	
Tokunagayusurika	Chironomidae, Diptera
Tokunagayusurika akamusi (Tokunaga)	
Tolucamyia	Sarcophagidae, Diptera
Tolucamyia schrameli (Dodge)	
Topomyia	Culicidae, Diptera
Topomyia apsarae Klein	
Topomyia auriceps Brug	
Topomyia danaraji Ramalingam	

Topomyia rausai Miyagi   **see** *Topomyia apsarae*
Topomyia sabahensis Ramalingam & Ramakrishna
Topomyia spathulirostris Edwards
Topomyia suchariti Miyagi & Toma
Topomyia yanbarensis Miyagi

Toxorhynchites   Culicidae, Diptera
Toxorhynchites acaudatus (Leicester)
Toxorhynchites amboinensis (Doleschall)
Toxorhynchites angustiplatus Evenhuis & Steffan
Toxorhynchites aurifluus (Edwards)   **see** *Toxorhynchites christophi aurifluus*
Toxorhynchites bengalensis Rosenberg & Evenhuis
Toxorhynchites brevipalpis Theobald
Toxorhynchites changbaiensis Su & Wang   **see** *Toxorhynchites christophi*
Toxorhynchites christophi aurifluus (Edwards)
Toxorhynchites christophi christophi (Porchinskiĭ)
Toxorhynchites coeruleus (Brug)
Toxorhynchites gravelyi (Edwards)
Toxorhynchites haemorrhoidalis (Fabricius)
Toxorhynchites indicus Evenhuis & Steffan
Toxorhynchites kempi (Edwards)
Toxorhynchites manicatus manicatus (Edwards)
Toxorhynchites manicatus yaeyamae Bohart
Toxorhynchites manicatus yamadai Ôuchi
Toxorhynchites moctezuma (Dyar & Knab)   **see** *Toxorhynchites theobaldi*
Toxorhynchites nepenthis (Dyar & Shannon)
Toxorhynchites ramalingami Evenhuis & Steffan
Toxorhynchites rutilus rutilus (Coquillett)
Toxorhynchites rutilus septentrionalis (Dyar & Knab)
Toxorhynchites splendens (Wiedemann)
Toxorhynchites theobaldi (Dyar & Knab)
Toxorhynchites towadensis (Matsumura)
Toxorhynchites yaeyamae Bohart   **see** *Toxorhynchites manicatus yaeyamae*

Tracheliastes   Lernaeopodidae, Copepoda
Tracheliastes maculatus (Kollar)

Tramea   Libellulidae, Odonata
Tramea abdominalis (Rambur)
Tramea limbata (Desjardins)

Triatoma   Reduviidae, Hemiptera
Triatoma barberi Usinger
Triatoma brasiliensis Neiva
Triatoma cavernicola Else & Cheong
Triatoma circummaculata (Stål)
Triatoma delpontei Romaña & Abalos
Triatoma dimidiata (Latreille)
Triatoma dispar Lent
Triatoma eratyrusiformis Del Ponte
Triatoma flavida Neiva
Triatoma gerstaeckeri (Stål)
Triatoma guasayana Wygodzinsky & Abalos
Triatoma infestans (Klug)
Triatoma lecticularia (Stål)
Triatoma longipennis Usinger
Triatoma maculata (Erichson)
Triatoma matogrossensis Leite & Barbosa
Triatoma mazzottii Usinger
Triatoma megista (Burmeister)   **see** *Panstrongylus megistus*
Triatoma melanocephala Neiva & Pinto
Triatoma pallidipennis (Stål)
Triatoma patagonica Del Ponte
Triatoma phyllosoma intermedia Usinger   **see** *Triatoma longipennis*

Triatoma phyllosoma pallidipennis (Stål) **see** *Triatoma pallidipennis*
Triatoma phyllosoma phyllosoma (Burmeister)
Triatoma platensis Neiva
Triatoma protracta (Uhler)
Triatoma pseudomaculata Corrêa & Spinola
Triatoma recurva (Stål)
Triatoma rubida rubida (Uhler)
Triatoma rubida uhleri (Neiva)
Triatoma rubrofasciata (DeGeer)
Triatoma rubrovaria (Blanchard)
Triatoma sanguisuga (LeConte)
Triatoma sordida (Stål)
Triatoma spinolai Porter
Triatoma vitticeps (Stål)

Trichiorhyssemus Scarabaeidae, Coleoptera
Trichiorhyssemus balthasari Bakovic
Trichiorhyssemus hirsutus (Clouet)
Trichiorhyssemus malkini Bakovic
Trichiorhyssemus samoanus Balthasar

Trichobius Streblidae, Diptera
Trichobius leionotus Wenzel
Trichobius major Coquillett

Trichoblatta Derocalymmidae, Dictyoptera
Trichoblatta nigra Shiraki
Trichoblatta pygmaea Karny

Trichodectes Trichodectidae, Phthiraptera
Trichodectes canis (DeGeer)
Trichodectes ermineae Hopkins
Trichodectes jacobi Eichler
Trichodectes melis (Fabricius)
Trichodectes mustelae (Schrank)
Trichodectes pilosus Giebel **see** *Werneckiella equi*
Trichodectes retusus Burmeister
Trichodectes subrostratus Burmeister **see** *Felicola subrostratus*

Trichoecius Myocoptidae, Acari
Trichoecius apodemi Fain, Munting & Lukoschus
Trichoecius clethrionomydis Portús & Gállego
Trichoecius gettingeri Fain et al.
Trichoecius pitymydis Portús & Gállego
Trichoecius tenax (Michael)

Trichogramma Trichogrammatidae, Hymenoptera
Trichogramma semblidis (Aurivillius)

Tricholipeurus **see** *Damalinia*
Tricholipeurus balanicus (Werneck) **see** *Damalinia cornuta*
Tricholipeurus lipeuroides (Mégnin) **see** *Damalinia lipeuroides*
Tricholipeurus parallelus (Osborn) **see** *Damalinia parallela*

Trichomalopsis Pteromalidae, Hymenoptera
Trichomalopsis dubius (Ashmead)

Trichonyssus Dermanyssidae, Acari
Trichonyssus australicus (Womersley)
Trichonyssus caputmedusae Domrow
Trichonyssus lukoschusi Micherdzinski & Domrow
Trichonyssus nixoni Micherdzinski & Domrow
Trichonyssus solivagus Domrow
Trichonyssus streetorum Micherdzinski & Domrow

Trichonyssus womersleyi Domrow

Trichopria
Trichopria lewisi Nixon

Diapriidae, Hymenoptera

Trichoprosopon
Trichoprosopon digitatum (Rondani)

Culicidae, Diptera

Trichopsyllopus

Acaridae, Acari

Trichopticoides
Trichopticoides decolor (Fallén)

see *Drymeia*
see *Drymeia vicana*

Trichoribates
Trichoribates novus (Sellnick)

Ceratozetidae, Acari

Trichosia
Trichosia pubescens Morgante

Sciaridae, Diptera

Trichouropoda
Trichouropoda orbicularis (Koch)

Uropodidae, Acari

Triglyphothrix
Triglyphothrix lanuginosa (Mayr)

see *Tetramorium*
see *Tetramorium lanuginosum*

Trigona
Trigona daemoniaca (Camargo)
Trigona mellicolor Packard

Apidae, Hymenoptera

Trinoton
Trinoton anserinum (Fabricius)
Trinoton querquedulae (Linnaeus)

Menoponidae, Phthiraptera

Triphleba
Triphleba lugubris (Meigen)

Phoridae, Diptera

Tripteroides
Tripteroides aranoides (Theobald)
Tripteroides bambusa bambusa (Yamada)

Culicidae, Diptera

Trischiza
Trischiza atricornis Ashmead

Figitidae, Hymenoptera

Trite

Salticidae, Araneae

Trixacarus
Trixacarus caviae Fain, Hovell & Hyatt

Sarcoptidae, Acari

Trombicula
Trombicula akamushi deliensis Walch
Trombicula autumnalis (Shaw)
Trombicula batatas (Linnaeus)
Trombicula dimolinae Audy
Trombicula knighti Radford
Trombicula multisetosa (Ewing)

Trombiculidae, Acari
see *Leptotrombidium deliense*
see *Neotrombicula autumnalis*
see *Eutrombicula batatas*
see *Rudnicula dimolinae*
see *Rudnicula knighti*
see *Eutrombicula multisetosa*

Trombiculindus
Trombiculindus lukoschusi Goff
Trombiculindus ochotonae Wang, Zhai & Chen
Trombiculindus quanzhouensis Liao, Li & Wang
Trombiculindus spinifoliatus Wang, Li & Tian

Trombiculidae, Acari

see *Leptotrombidium ochotonae*
see *Leptotrombidium quanzhouensis*
see *Leptotrombidium spinifoliatum*

Trombigastia
Trombigastia guangdongensis Goff *et al.*

Trombiculidae, Acari

Tropiconabis	**see** *Nabis*
Tropisternus	Hydrophilidae, Coleoptera
Tropisternus lateralis (Fabricius)	
Trouessartia	Trouessartiidae, Acari
Trouessartia amplivasa Gaud & Atyeo	
Trouessartia appendiculata (Berlese)	
Trouessartia botulifera Gaud & Atyeo	
Trouessartia crucifera Gaud	
Trouessartia gabonica Gaud & Atyeo	
Trouessartia gladifera Gaud & Atyeo	
Trouessartia juliettae Gaud & Atyeo	
Trouessartia microcaudata Mironov	
Trouessartia minutipes (Berlese)	
Trouessartia paludicolae Gaud & Atyeo	
Trouessartia quarta Gaud & Atyeo	
Trouessartia ripariae Mironov	
Trox	Trogidae, Coleoptera
Trox aino Makane & Tsukamoto	
Trox cadaverinus Illiger	
Trox fabricii Reiche	
Trox hispidus (Pontoppidan)	
Trox scaber (Linnaeus)	
Trox setifer Waterhouse	
Trypocalliphora	Callophoridae, Diptera
Trypocalliphora braueri (Hendel)	
Trypocopris	**see** *Geotrupes*
Trypocopris pyrenaeus (Charpentier)	**see** *Geotrupes pyrenaeus*
Trypocopris vernalis (Linnaeus)	**see** *Geotrupes vernalis*
Trypoxylon	Sphecidae, Hymenoptera
Trypoxylon politum Say	
Tunga	Tungidae, Siphonaptera
Tunga caecigena Jordan & Rothschild	
Tunga monositus Barnes & Radovsky	
Tunga penetrans (Linnaeus)	
Tunga terasma Jordan	
Turdinirmus	**see** *Brueelia*
Turdinirmus merulensis Denny	**see** *Brueelia merulensis*
Tuxophorus	Caligidae, Copepoda
Twinnia	Simuliidae, Diptera
Twinnia cannibora Ono	
Tydeus	Tydeidae, Acari
Tydeus interruptus Thor	
Tylotropidius	Acrididae, Orthoptera
Tylotropidius patagiatus (Karsch)	
Typhaeus	Geotrupidae, Coleoptera
Typhaeus typhoeus (Linnaeus)	
Typhloceras	Hystrichopsyllidae, Siphonaptera
Typhloceras poppei Wagner	

Typhlomyophthirus
Typhlomyophthirus bifoliatus Chin

Haematopinoididae, Phthiraptera

Typhlomyopsyllus
Typhlomyopsyllus esinus Liu, Shi & Liu

Leptopsyllidae, Siphonaptera

Tyroglyphus
Tyroglyphus farinae (Linnaeus)

see *Acarus*
see *Acarus siro*

Tyrophagus
Tyrophagus castellanii (Hirst)
Tyrophagus longior (Gervais)
Tyrophagus putrescentiae (Schrank)
Tyrophagus similis Volgin

Acaridae, Acari
see *Tyrophagus putrescentiae*

Tyrrellia
Tyrrellia circularis Koenike

Limnesiidae, Acari

U

Uchida
Uchida phasiani Modrzejewska & Złotorzycka

Menoponidae, Phthiraptera

Unionicola
Unionicola abnormipes (Wolcott)
Unionicola fossulata (Koenike)
Unionicola serrata (Wolcott)

Unionicolidae, Acari

Uranotaenia
Uranotaenia bilineata bilineata Theobald
Uranotaenia bilineata fraseri Edwards
Uranotaenia caliginosa Philip
Uranotaenia gouldi Peyton & Klein
Uranotaenia henrardi Edwards
Uranotaenia hirsutifemora Peters
Uranotaenia lowii Theobald
Uranotaenia machadoi Ramos
Uranotaenia maculipleura Leicester
Uranotaenia mayeri Edwards
Uranotaenia ohamai Tanaka, Mizusawa & Saugstad
Uranotaenia sapphirina (Osten Sacken)
Uranotaenia unguiculata Edwards

Culicidae, Diptera

Urodacus
Urodacus yaschenkoi (Birula)

Scorpionidae, Scorpiones

Urolepis
Urolepis rufipes (Ashmead)

Pteromalidae, Hymenoptera

Uropsylla
Uropsylla tasmanica Rothschild

Pygiopsyllidae, Siphonaptera

Uroseius
Uroseius acuminatus (Koch)

Uropodidae, Acari

Uroxys
Uroxys transversifrons Howden & Gill

Scarabaeidae, Coleoptera

Uvarus

Dytiscidae, Coleoptera

V

Vaejovis
Vaejovis wupatkiensis Stahnke

Vaejovidae, Scorpiones
see *Serradigitus wupatkiensis*

Vargatula	Trombiculidae, Acari
Vargatula somaliensis Goff	
Veigaia	Veigaiidae, Acari
Veigaia kochi (Trägårdh)	
Veigaia nemorensis (Koch)	
Vermipsylla	Vermipsyllidae, Siphonaptera
Verrallina	**see** *Aedes*
Vespa	Vespidae, Hymenoptera
Vespa analis Fabricius	
Vespa basalis F. Smith	
Vespa cincta Fabricius	**see** *Vespa tropica*
Vespa crabro Linnaeus	
Vespa germanica Fabricius	**see** *Vespula germanica*
Vespa luctuosa Saussure	
Vespa mandarinia japonica Radoszkowski	
Vespa mandarinia mandarinia F. Smith	
Vespa orientalis orientalis Linnaeus	
Vespa orientalis somalica Soika	
Vespa simillima simillima F. Smith	
Vespa simillima xanthoptera Cameron	
Vespa tropica deusta Lepeletier	
Vespa tropica tropica (Linnaeus)	
Vespa velutina pruthii van der Vecht	
Vespa xanthoptera Cameron	**see** *Vespa simillima xanthoptera*
Vespula	Vespidae, Hymenoptera
Vespula acadica (Sladen)	
Vespula arenaria (Fabricius)	**see** *Dolichovespula arenaria*
Vespula atropilosa (Sladen)	
Vespula austriaca (Panzer)	
Vespula consobrina (Saussure)	
Vespula flaviceps flaviceps (F. Smith)	
Vespula flaviceps lewisii (Cameron)	
Vespula flavopilosa Jacobson	
Vespula germanica (Fabricius)	
Vespula lewisii (Cameron)	**see** *Vespula flaviceps lewisii*
Vespula maculata (Linnaeus)	**see** *Dolichovespula maculata*
Vespula maculifrons (Buysson)	
Vespula pensylvanica (Saussure)	
Vespula squamosa (Drury)	
Vespula vidua (Saussure)	
Vespula vulgaris (Linnaeus)	
Viedebanttia	Acaridae, Acari
Viedebanttia diamanus Fain & Schwan	
Viedebanttia schmitzi Oudemans	
W	
Walchia	Trombiculidae, Acari
Walchia americana Ewing	
Walchia masoni (Asanuma & Saito)	
Walchia ogati Sasa & Tetramura	
Walchia xishaensis Zhao, Tang & Mo	
Walchiella	Trombiculidae, Acari
Walchiella oudemansi katangladensis Kolebinova	
Walchiella oudemansi oudemansi (Walch)	
Walchiella philippinensis Kolebinova	

Walzia
Walzia australica Womersley

Anystidae, Acari

Wasmannia
Wasmannia auropunctata (Roger)

Formicidae, Hymenoptera
see *Ochetomyrmex auropunctatus*

Werneckiella
Werneckiella equi asini Eichler
Werneckiella equi equi (Denny)

Trichodectidae, Phthiraptera

Whartonia
Whartonia nudosetosa (Wharton)

Trombiculidae, Acari

Wilhelmia
Wilhelmia equina (Linnaeus)
Wilhelmia lineata (Meigen)
Wilhelmia mediterranea (Puri)
Wilhelmia pseudequina (Séguy)
Wilhelmia talassica Yankovskiĭ
Wilhelmia veltistshevi (Rubtsov)

see *Simulium*
see *Simulium equinum*
see *Simulium lineatum*
see *Simulium pseudequinum*
see *Simulium pseudequinum*
see *Simulium talassicum*
see *Simulium veltistshevi*

Wohlfahrtia
Wohlfahrtia magnifica (Schiner)
Wohlfahrtia nuba (Wiedemann)
Wohlfahrtia opaca (Coquillett)
Wohlfahrtia vigil opaca (Coquillett)
Wohlfahrtia vigil vigil (Walker)

Sarcophagidae, Diptera

see *Wohlfahrtia vigil opaca*

Wyeomyia
Wyeomyia aporonoma Dyar & Knab
Wyeomyia leucostigma Lutz
Wyeomyia mitchellii (Theobald)
Wyeomyia smithii (Coquillett)
Wyeomyia vanduzeei Dyar & Knab

Culicidae, Diptera

Wynnowenia
Wynnowenia distinctus Boxshall

Hatschekiidae, Copepoda

X
Xeniaria
Xeniaria jacobsoni (Burr)

Arixeniidae, Dermaptera

Xeniarioptes
Xeniarioptes scutellatus Fain & Lukoschus

Rosensteiniidae, Acari

Xenillus
Xenillus itascensis Freeman & Woolley

Liacaridae, Acari

Xenocalliphora
Xenocalliphora hortona (Walker)

Calliphoridae, Diptera

Xenopsylla
Xenopsylla astia Rothschild
Xenopsylla bantorum Jordan
Xenopsylla blanci Smit
Xenopsylla brasiliensis (Baker)
Xenopsylla cheopis (Rothschild)
Xenopsylla conformis (Wagner)
Xenopsylla cunicularis Smit
Xenopsylla eridos (Rothschild)
Xenopsylla gerbilli caspica Ioff
Xenopsylla gerbilli gerbilli (Wagner)
Xenopsylla gerbilli minax Jordan
Xenopsylla guancha Beaucournu, Alcover & Launay

Pulicidae, Siphonaptera

Xenopsylla hirtipes Rothschild
Xenopsylla nuttali Ioff
Xenopsylla piriei Ingram
Xenopsylla skrjabini Ioff
Xenopsylla vexabilis Jordan

Xenoryctes   Glycyphagidae, Acari
Xenoryctes krameri (Michael)

Xeromegachile   see *Megachile*

Xyleutes   Cossidae, Lepidoptera
Xyleutes leucomochla Turner

Xylocopa   Apidae, Hymenoptera
Xylocopa tenuiscapa (Westwood)
Xylocopa violacea (Linnaeus)

Xysticus   Thomisidae, Araneae
Xysticus cristatus (Clerck)

Youngaphodius   see *Aphodius*

Y
Yunkeracarus   Gastronyssidae, Acari
Yunkeracarus faini apodemi Fain *et al.*
Yunkeracarus microti Smith *et al.*
Yunkeracarus muris Fain

Z
Zachvatkinia   Avenzoariidae, Acari
Zachvatkinia hydrobatidii Dubinin
Zachvatkinia stercorarii Dubinin

Zelurus   Reduviidae, Hemiptera
Zelurus femoralis femoralis (Stål)
Zelurus femoralis intermedius Lent & Wygodzinsky

Zerconopsis   Ascidae, Acari
Zerconopsis remiger (Kramer)

Zeteticontus   see *Cerchysiella*

Zeugnomyia   Culicidae, Diptera

Zibethacarus   Acaridae, Acari
Zibethacarus ondatrae (Rupeš & Whitaker)

Zlotorzyckella   Goniodidae, Phthiraptera
Zlotorzyckella colchici (Denny)

Zosimus   Xanthidae, Decapoda
Zosimus aeneus (Linnaeus)

Zumptiella   Halarachnidae, Acari
Zumptiella tamias Fain, Lukoschus & Whitaker

Zumptrombicula   Trombiculidae, Acari
Zumptrombicula misonnei Goff

Zuskamira   Sepsidae, Diptera
Zuskamira inexpectata Pont

Zygochelifer   Avenzoariidae, Acari

Zygochelifer psalidurus (Trouessart)

Zygoribatula   Oribatulidae, Acari
Zygoribatula lata Hammer

MICROORGANISMS

The following are the most important microorganisms (except viruses) encountered in medical and veterinary entomology. The family and order/another taxon of each is given to the right, except where a synonym is present.

A

Actinomyces
Actinomyces pyogenes (Glage)

Actinomycetaceae, Actinomycetales

Adelina
Adelina riouxi Levine

Adeleidae, Eucoccidiorida

Amblyospora
Amblyospora benigna Kellen & Wills
Amblyospora californica (Kellen & Lipa)
Amblyospora campbelli (Kellen & Wills)
Amblyospora capillata (Larsson)
Amblyospora culicis Toguebaye & Marchland
Amblyospora dyxenoides Sweeney *et al.*
Amblyospora egypti Darwish & Canning
Amblyospora gigantea (Kellen & Wills)
Amblyospora indicola Vávra *et al.*
Amblyospora inimica (Kellen & Wills)
Amblyospora kadunae Weiser & Prasertphon
Amblyospora minuta (Kudo)
Amblyospora mojingensis Hazard & Oldacre
Amblyospora nigeriana Weiser & Prasertphon
Amblyospora noxia (Kellen & Wills)
Amblyospora opacita (Kudo)
Amblyospora pinensis Kettle & Piper

Thelohaniidae, Microsporida

Anaplasma
Anaplasma centrale (Theiler)
Anaplasma marginale Theiler
Anaplasma ovis Lestoquard

Anaplasmataceae, Rickettsiales

Ascocystis
Ascocystis chagasi (Adler & Mayrink)
Ascocystis culicis (Ross)
Ascocystis taiwanensis Lien & Levine

Lecudinidae, Eugregarinorida
see *Ascogregarina chagasi*
see *Ascogregarina culicis*
see *Ascogregarina taiwanensis*

Ascogregarina
Ascogregarina barretti (Vavra)
Ascogregarina chagasi (Adler & Mayrink)
Ascogregarina clarki (Sanders & Poinar)
Ascogregarina culicis (Ross)
Ascogregarina geniculati Munstermann & Levine
Ascogregarina taiwanensis (Lien & Levine)

Lecudinidae, Eugregarinorida

B

Babesia
Babesia argentina Lignières
Babesia bigemina Smith & Killbourne
Babesia bovis (Babès)
Babesia caballi (Nuttall)
Babesia canis (Piana & Galli-Valerio)
Babesia divergens (M'Fadyean & Stockman)
Babesia equi Laveran
Babesia felis Davis
Babesia gibsoni (Patton)

Babesiidae, Piroplasmorida
see *Babesia bovis*

Babesia herpailuri Dennig
Babesia jakimovi Nikolskii *et al.*
Babesia major (Sergent *et al.*)
Babesia microti (Coles)
Babesia motasi Wenyon
Babesia occultans Gray & de Vos
Babesia ovata Minami & Ishihara
Babesia ovis (Babés)
Babesia perroncitoi (Cerruti)
Babesia trautmanni (Knuth & Du Toit)
Babesia vesperuginis (Dionisi)
Babesia yakimovi Sprinholz–Schmidt

Bacillus   Bacillaceae, Eubacteriales
Bacillus sphaericus Meyer & Neide
Bacillus thuringiensis aizawai Bonnefoi & de Barjac
Bacillus thuringiensis entomocidus Steinhaus
Bacillus thuringiensis finitimus Heimpel & Angus
Bacillus thuringiensis galleriae Isakova
Bacillus thuringiensis israelensis de Barjac
Bacillus thuringiensis kenyae Norris & Burgess
Bacillus thuringiensis kurstaki Kurstak
Bacillus thuringiensis morrisoni Norris
Bacillus thuringiensis tenebrionis Krieg *et al.*
Bacillus thuringiensis thuringiensis Berliner

Bartonella   Bartonellaceae, Rickettsiales
Bartonella bacilliformis (Strong *et al.*)

Blastocrithidia   Trypanosomatidae, Kinetoplastorida
Blastocrithidia triatomae (Cerisola *et al.*)

Bohuslavia   Thelohaniidae, Microsporida
Bohuslavia asterias (Weiser)

Borrelia   Spirochaetaceae, Spirochaetales
Borrelia anserina (Sakharoff)
Borrelia burgdorferi Johnson *et al.*
Borrelia duttoni (Novy & Knapp)
Borrelia hermsii (Davis)
Borrelia persica (Dschunkowsky)
Borrelia recurrentis (Lebert)
Borrelia sogdiana (Nicolle & Anderson)   **see** *Borrelia persica*

Brucella   Brucellaceae, –
Brucella canis Carmichael & Bruner

Burenella   Burenellidae, Microsporida

C
Campylobacter   Spirillaceae, Gracilicutes
Campylobacter fetus (Smith & Taylor)
Campylobacter fetus jejuni Jones *et al.*   **see** *Campylobacter jejuni*
Campylobacter jejuni (Jones *et al.*)

Caudospora   Caudosporidae, Microsporida
Caudospora simulii Weiser

Chainia   **see** *Streptomyces*

Chapmanium   Thelohaniidae, Microsporida
Chapmanium dispersus Larsson   **see** *Napamichum dispersus*

Citrobacter	Enterobacteriaceae, Gracilicutes
Citrobacter diversus (Burkey)	
Citrobacter freundii (Braak)	
Cowdria	Rickettsiaceae, Rickettsiales
Cowdria ruminantium (Cowdry)	
Corynebacterium	Actinomycetaceae, Actinomycetales
Corynebacterium pyogenes (Glage)	**see** *Actinomyces pyogenes*
Coxiella	Rickettsiaceae, Rickettsiales
Coxiella burnetii (Derrick)	
Crithidia	Trypanosomatidae, Kinetoplastorida
Crithidia bombi Lipa & Triggiani	
Crithidia guilhermei Soares *et al.*	
Culicospora	Culicosporidae, Microsporida
Culicospora magna (Kudo)	
Culicosporella	Caudosporidae, Microsporida
Culicosporella lunata (Hazard & Savage)	
Cylindrospora	Thelohaniidae, Microsporida
Cylindrospora chironomi Issi & Voronin	
Cytauxzoon	Theileriidae, Piroplasmorida
Cytauxzoon felis Kier	**see** *Theileria felis*
Cytauxzoon taurotragi Martin & Brocklesby	**see** *Theileria taurotragi*
Cytoecetes	Rickettsiaceae, Rickettsiales
Cytoecetes phagocytophila (Foggie)	**see** *Ehrlichia phagocytophila*

D

Dermatophilus	Dermatophilaceae, Actinomycetales
Dermatophilus congolensis Van Saceghem	
Duboscqia	Duboscqiidae, Microsporida
Duboscqia aediphaga Kettle & Piper	

E

Ehrlichia	Rickettsiaceae, Rickettsiales
Ehrlichia canis (Donatien & Lestoquard)	
Ehrlichia phagocytophila (Foggie)	
Endolimax	Endamoebidae, Amoebida
Endolimax nana (Wenyon & O'Connor)	
Entamoeba	Endamoebidae, Amoebida
Entamoeba coli (Grassi)	
Entamoeba histolytica Schaudinn	
Eperythrozoon	Anaplasmataceae, Rickettsiales
Eperythrozoon ovis (Neitz *et al.*)	
Evlachovaia	Burenellidae, Microsporida
Evlachovaia chironomi Voronin & Issi	

F

Flabelliforma	– , Microsporida
Flabelliforma montana Canning *et al.*	

Francisella *Francisella novicida* (Larson *et al.*) *Francisella philomiragia* (Jensen *et al.*) *Francisella tularensis* (McCoy & Chapin)	Enterobacteriaceae, Gracilicutes **see** *Francisella tularensis*
Fusobacterium *Fusobacterium necrophorum* (Flügge)	Bacteroidaceae, Gracilicutes

G

Giardia *Giardia lamblia* (Stiles)	Hexamitidae, Diplomonadorida
Gonderia *Gonderia bovis* Neitz *Gonderia orientalis* Yakimoff & Soudatschenkoff	Theileriidae, Piroplasmorida **see** *Theileria bovis* **see** *Theileria orientalis*

H

Haematoxenus *Haematoxenus veliferus* Uilenberg	Theileriidae, Piroplasmorida **see** *Theileria velifera*
Haemobartonella *Haemobartonella canis* (Kikuth)	Anaplasmataceae, Rickettsiales
Haemophilus	Pasteurellaceae, Gracilicutes
Haemoproteus *Haemoproteus columbae* Kruse *Haemoproteus meleagridis* Levine	Haemoproteidae, Haemospororida
Hafnia	Enterobacteriaceae, Gracilicutes
Hazardia *Hazardia milleri* (Hazard & Fukuda)	Culicosporidae, Microsporida
Helmichia *Helmichia aggregata* Larsson	Thelohaniidae, Microsporida
Hepatocystis *Hepatocystis levinei* Landau *et al.*	Haemoproteidae, Haemospororida
Hepatozoon *Hepatozoon canis* (James) *Hepatozoon erhardovae* Krampitz	Hepatozooidae, Eucoccidiorida
Herpetomonas *Herpetomonas dedonderi* Dedet *et al.*	Trypanosomatidae, Kinetoplastorida
Hyalinocysta *Hyalinocysta expilatoria* Larsson	Thelohaniidae, Microsporida

L

Lactobacillus *Lactobacillus plantarum* (Orla–Jensen)	Lactobacillaceae, Eubacteriales
Leidyana *Leidyana guttiventrisa* Sarker	Leidyanidae, Eugregarinorida
Leishmania *Leishmania adleri* (Heisch) *Leishmania aethiopica* Bray, Ashford & Bray *Leishmania agamae* (David) *Leishmania amazonensis* Lainson & Shaw *Leishmania arabica* Peters, Elbihari & Evans	Trypanosomatidae, Kinetoplastorida **see** *Sauroleishmania agamae*

Leishmania archibaldi Castellani & Chalmers
Leishmania aristidesi Lainson & Shaw
Leishmania braziliensis Vianna
Leishmania braziliensis guyanensis Floch — **see** *Leishmania guyanensis*
Leishmania braziliensis panamensis Lainson & Shaw — **see** *Leishmania panamensis*
Leishmania braziliensis peruviana Valez — **see** *Leishmania peruviana*
Leishmania ceramodactyli (Adler & Theodor) — **see** *Sauroleishmania ceramodactyli*
Leishmania chagasi Cunha & Chagas — **see** *Leishmania infantum chagasi*
Leishmania chamaeleonis (Wenyon) — **see** *Sauroleishmania chamaeleonis*
Leishmania davidi (Strong) — **see** *Sauroleishmania davidi*
Leishmania deanei Lainson & Shaw
Leishmania donovani (Laveran & Mesnil)
Leishmania donovani archibaldi Castellani & Chalmers — **see** *Leishmania archibaldi*
Leishmania donovani chagasi Cunha & Chagas — **see** *Leishmania infantum chagasi*
Leishmania donovani infantum Nicolle — **see** *Leishmania infantum infantum*
Leishmania aethiopica Bray *et al.*
Leishmania enriettii Muniz & Medina
Leishmania garnhami Scorza *et al.*
Leishmania gerbilli Wang *et al.*
Leishmania guliki Ovezmukhammedov & Saf'yanova — **see** *Sauroleishmania guliki*
Leishmania guyanensis Floch
Leishmania gymnodactyli (Khodukin & Sofiev) — **see** *Sauroleishmania gymnodactyli*
Leishmania hemidactyli (Mackie *et al.*) — **see** *Sauroleishmania hemidactyli*
Leishmania henrici (Léger) — **see** *Sauroleishmania henrici*
Leishmania herreri Zeledón *et al.*
Leishmania hertigi Herrer
Leishmania hertigi deanei Lainson & Shaw — **see** *Leishmania deanei*
Leishmania hoogstraali (McMillan) — **see** *Sauroleishmania hoogstraali*
Leishmania infantum chagasi Cunha & Chagas
Leishmania infantum infantum Nicolle
Leishmania killicki Rioux *et al.*
Leishmania lainsoni Silveira *et al.*
Leishmania major Yakimoff & Schokhor
Leishmania mexicana Biagi
Leishmania mexicana amazonensis Lainson & Shaw — **see** *Leishmania amazonensis*
Leishmania mexicana aristidesi Lainson & Shaw — **see** *Leishmania aristidesi*
Leishmania mexicana enriettii Muniz & Medina — **see** *Leishmania enriettii*
Leishmania mexicana garnhami Scorza *et al.* — **see** *Leishmania garnhami*
Leishmania mexicana pifanoi Medina & Romero — **see** *Leishmania pifanoi*
Leishmania mexicana venezuelensis Bonfante-Garrido — **see** *Leishmania venezuelensis*
Leishmania minor Yakimoff & Schokhor
Leishmania naiffi Lainson & Shaw
Leishmania nicollei (Khodukin & Sofiev) — **see** *Sauroleishmania nicollei*
Leishmania panamensis Lainson & Shaw
Leishmania peruviana Velez
Leishmania pifanoi Medina & Romero
Leishmania senegalensis (Ranque) — **see** *Sauroleishmania senegalensis*
Leishmania shawi Lainson *et al.*
Leishmania sofieffi (Markov *et al.*) — **see** *Sauroleishmania sofieffi*
Leishmania tarentolae (Wenyon) — **see** *Sauroleishmania tarentolae*
Leishmania tropica (Wright)
Leishmania tropica aethiopica Bray *et al.* — **see** *Leishmania aethiopica*
Leishmania tropica major Yakimoff & Schokhor — **see** *Leishmania major*
Leishmania tropica minor Yakimoff & Schokhor — **see** *Leishmania minor*
Leishmania turanica Strelkova *et al.*
Leishmania venezuelensis Bonfante-Garrido
Leishmania zmeevi Andrushko & Markov — **see** *Sauroleishmania zmeevi*

Leptospira — Leptospiraceae, Spirochaetales
Leptospira interrogans (Stimson)

Leucocytozoon — Plasmodiidae, Haemospororida
Leucocytozoon caulleryi Mathis & Léger
Leucocytozoon simondi Mathis & Léger

Leucocytozoon smithi (Laveran & Lucet)

M

Moraxella
Moraxella bovis (Hauduroy)
Moraxella ovis (Lindqvist)

Neisseriaceae, Gracilicutes

Morganella
Morganella morganii (Winslow *et al.*)

Enterobacteriaceae, Gracilicutes

N

Napamichum
Napamichum dispersus (Larsson)

Thelohaniidae, Microsporida

Nocardia
Nocardia brasiliensis (Lindenberg)

Nocardiaceae, Actinomycetales

Nosema
Nosema algerae Vávra & Undeen
Nosema hybomitrae

Nosematidae, Microsporida

see *Vairimorpha hybomitrae*

O

Octosporea
Octosporea muscaedomesticae Flu

– , Microsporida

P

Parathelohania
Parathelohania anophelis (Kudo)
Parathelohania arabiensis Weiser & Ibrahim
Parathelohania obesa (Kudo)

Amblyosporidae, Microsporida

Pasteurella
Pasteurella multocida (Lehmann & Neumann)
Pasteurella pestis Lehmann & Neumann
Pasteurella tularensis (McCoy & Chapin)

Enterobacteriaceae, Gracilicutes

see *Yersinia pestis*
see *Francisella tularensis*

Pegmatheca
Pegmatheca simulii Hazard & Oldacre

Thelohaniidae, Microsporida

Peptostreptococcus
Peptostreptococcus indolicus (Christiansen)

Peptococcaceae, Firmicutes

Pilosporella
Pilosporella chapmani Hazard & Oldacre

Thelohaniidae, Microsporida

Plasmodium
Plasmodium berghei Vincke & Lips
Plasmodium chabaudi Landau
Plasmodium cynomolgi Mayer
Plasmodium falciparum (Welch)
Plasmodium fieldi Eyles *et al.*
Plasmodium floridense Thomson & Huff
Plasmodium gallinaceum Brumpt
Plasmodium hermani Telford & Forrester
Plasmodium inui Halberstädter & Prowazek
Plasmodium kempi Christensen *et al.*
Plasmodium knowlesi Sinton & Mulligan
Plasmodium malariae (Laveran)
Plasmodium mexicanum Thompson & Huff
Plasmodium ovale Stephens
Plasmodium traguli Garnham & Edeson
Plasmodium vivax (Grassi & Feletti)
Plasmodium yoelii yoelii (Landau & Killick-Kendrick)

Plasmodiidae, Haemospororida

Plasmodium yoelii nigeriensis (Killick-Kendrick)

Pleistophora   Pleistophoridae, Microsporida
Pleistophora caecorum (Chapman & Kellen)
Pleistophora chaobori (Rapsch)   **see** *Systenostrema corethrae*
Pleistophora lutzi Sprague
Pleistophora multispora (Strickland)
Pleistophora simulii Lutz & Splendore   **see** *Polydispyrenia simulii*

Polychromophilus   Plasmodiidae, Haemospororida
Polychromophilus murinus Dionisi

Polydispyrenia   Pleistophoridae, Pleistophoridida
Polydispyrenia simulii (Lutz & Splendore)

Proteus   Enterobacteriaceae, Gracilicutes
Proteus mirabilis Hauser
Proteus morganii (Winslow *et al.*)   **see** *Morganella morganii*
Proteus rettgeri (Hadley *et al.*)   **see** *Providencia rettgeri*
Proteus vulgaris Hauser

Providencia   Enterobacteriaceae, Gracilicutes
Providencia rettgeri (Hadley *et al.*)
Providencia stuartii (Buttiaux *et al.*)

Pseudomonas   Pseudomonadaceae, Pseudomonadales
Pseudomonas aeruginosa (Schroeter)
Pseudomonas fluorescens (Trevisan)
Pseudomonas maltophilia Hugh
Pseudomonas stutzeri (Lehmann & Neumann)

R
Rickettsia   Rickettsiaceae, Rickettsiales
Rickettsia akari Hübner *et al.*
Rickettsia australis Philip
Rickettsia canada McKiel *et al.*
Rickettsia conorii Brumpt
Rickettsia montana (Lackmann *et al.*)
Rickettsia mooseri Monteiro   **see** *Rickettsia typhi*
Rickettsia parkeri Lackmann *et al.*
Rickettsia prowazekii Da Rocha-Lima
Rickettsia quintana Schminke   **see** *Rochalimaea quintana*
Rickettsia rhipicephali (Burgdorfer *et al.*)
Rickettsia rickettsii (Wolbach)
Rickettsia sibirica Zdrodovskii
Rickettsia slovaca (Úrvölgyi & Brezina)
Rickettsia tsutsugamushi (Hayashi)
Rickettsia typhi (Wolbach & Todd)

Rochalimaea   Rickettsiaceae, Rickettsiales
Rochalimaea quintana (Schminke)

S
Salmonella   Enterobacteriaceae, Gracilicutes
Salmonella paratyphi-A (Brion & Kayser)
Salmonella schoetmuelleri (Winslow *et al.*)
Salmonella typhi (Schroeter)

Sauroleishmania   Trypanosomatidae, Kinetoplastorida
Sauroleishmania adleri (Heisch)
Sauroleishmania agamae (David)
Sauroleishmania ceramodactyli (Adler & Theodor)
Sauroleishmania chamaeleonis (Wenyon)

Sauroleishmania davidae (Strong)
Sauroleishmania guliki (Ovezmukhammedov & Saf'yanova)
Sauroleishmania gymnodactyli (Khodukin & Sofiev)
Sauroleishmania hemidactyli (Mackie *et al.*)
Sauroleishmania henrici (Léger)
Sauroleishmania hoogstraali (McMillan)
Sauroleishmania nicollei (Khodukin & Sofiev)
Sauroleishmania senegalensis (Ranque)
Sauroleishmania sofieffi (Markov *et al.*)
Sauroleishmania tarentolae (Wenyon)
Sauroleishmania zmeevi (Andrushko & Markov)

Schellackia   Lankesterellidae, Eucoccidiorida
Schellackia golvani Rogier & Landau
Schellackia occidentalis Bonoris & Ball

Serratia   Enterobacteriaceae, Gracilicutes
Serratia marcescens Bizio

Shigella   Enterobacteriaceae, Gracilicutes
Shigella flexneri Castellani & Chalmers

Staphylococcus   Micrococcaceae, Gracilicutes
Staphylococcus aureus Rosenbach
Staphylococcus epidermidis (Winslow & Winslow)
Staphylococcus hyicus (Sompolinsky)
Staphylococcus intermedius Hajek

Streptomyces   Streptomycetaceae, Actinomycetales
Streptomyces anulatus (Beijerinck)
Streptomyces avermitilis
Streptomyces griseus (Krainsky)   **see** *Streptomyces anulatus*

Systenostrema   Thelohaniidae, Microsporida
Systenostrema corethrae Schuberg & Rodriquez

T
Tetrahymena   Tetrahymenidae, Hymenostomatida
Tetrahymena pyriformis Ehrenberg
Tetrahymena rotunda Lynn *et al.*

Theileria   Theileriidae, Piroplasmorida
Theileria annulata (Dschunkowsky & Luhs)
Theileria bovis (Neitz)
Theileria buffeli Neveu-Lemaire
Theileria camelensis Yakimoff
Theileria cervi (Bettencourt *et al.*)
Theileria felis (Kier)
Theileria hirci Dschunkowsky & Urodschevich
Theileria lawrencei (Neitz)
Theileria mutans (Theiler)
Theileria orientalis (Yakimoff & Soudatschenkoff)
Theileria ovis Rodhain
Theileria paramelis Mackerras
Theileria parva (Theiler)
Theileria parva bovis (Neitz)   **see** *Theileria bovis*
Theileria parva lawrencei (Neitz)   **see** *Theileria lawrencei*
Theileria recondita Lestoquard
Theileria sergenti Yakimoff & Dekhtereff
Theileria tarandi Kertzelli
Theileria taurotragi (Martin & Brocklesby)
Theileria velifera (Uilenberg)

Thelohania	Thelohaniidae, Microsporida
Thelohania asterias Weiser	**see** *Bohuslavia asterias*
Thelohania capillata Larsson	**see** *Amblyospora capillata*
Thelohania corethrae Schuberg & Rodriquez	**see** *Systenostrema corethrae*
Thelohania fibrata (Strickland)	
Thelohania varians (Léger)	
Treponema	Spirochaetaceae, Spirochaetales
Treponema pertenue Castellani	
Trichomonas	Trichomonadidae, Trichomonadorida
Trichomonas hominis (Davaine)	
Trypanosoma	Trypanosomatidae, Kinetoplastorida
Trypanosoma brucei brucei Plimmer & Bradford	
Trypanosoma brucei gambiense Dutton	**see** *Trypanosoma gambiense*
Trypanosoma brucei rhodesiense Stephens & Fantham	**see** *Trypanosoma rhodesiense*
Trypanosoma chabaudi Chandenier *et al.*	
Trypanosoma congolense Broden	
Trypanosoma conorhini (Donovan)	
Trypanosoma corvi Stephens & Christophers	
Trypanosoma cruzi (Chagas)	
Trypanosoma dionisii Bettencourt & França	
Trypanosoma equinum Voges	
Trypanosoma evansi (Steel)	
Trypanosoma gambiense Dutton	
Trypanosoma hedricki Bower & Woo	
Trypanosoma hippicum Darling	
Trypanosoma incertum Pittaluga	
Trypanosoma leschenaulti Robertson	
Trypanosoma melophagium (Flu)	
Trypanosoma musculi Kendall	
Trypanosoma myoti Bower & Woo	
Trypanosoma rangeli Téjera	
Trypanosoma rhodesiense Stephens & Fantham	
Trypanosoma simiae Bruce *et al.*	**see** *Trypanosoma congolense*
Trypanosoma theileri Laveran	
Trypanosoma vespertilionis (Battaglia)	
Trypanosoma vivax Ziemann	

V

Vairimorpha	Burenellidae, Microsporida
Vairimorpha hybomitrae	
Vairimorpha invictae Jouvenaz & Ellis	
Vavraia	Pleistophoridae, Microsporida
Vavraia culicis (Weiser)	
Vibrio	Vibrionaceae, Gracilicutes
Vibrio comma (Schroeter)	

X

Xenorhabdus	Enterobacteriaceae, Gracilicutes
Xenorhabdus luminescens Thomas & Poinar	
Xenorhabdus nematophilus (Poinar & Thomas)	

W

Wolbachia	Rickettsiaceae, Rickettsiales
Wolbachia pipientis Hertig	

Y

Yersinia	Enterobacteriaceae, Gracilicutes
Yersinia enterocolitica (Schleifstein & Coleman)	

Yersinia pestis (Lehmann & Neumann)
Yersinia philomiragia Jensen *et al.*   **see** *Francisella philomiragia*
Yersinia pseudotuberculosis pestis (Lehmann & Neumann) **see** *Yersinia pestis*

VIRUSES

The following are the most common viruses transmitted directly or indirectly by arthropods (not all are arboviruses). Where known, the genus and family (or synonym) of each is given to the right.

A

African horse sickness virus	*Orbivirus*, Reoviridae
African swine fever virus	– , –
Akabane virus	*Bunyavirus*, Bunyaviridae
Avalon virus	*Nairovirus*, Bunyaviridae
Avian influenzavirus	*Influenzavirus type A*, Orthomyxoviridae

B

Bhanja virus	*Bunyavirus*, Bunyaviridae
Biphasic milk fever virus	**see** Central European encephalitis virus
Bluetongue virus	*Orbivirus*, Reoviridae
Bovine diarrhoea virus	*Pestivirus*, Togaviridae
Bovine ephemeral fever virus	*Lyssavirus*, Rhabdoviridae
Bovine leukaemia virus	*Type C Oncovirus*, Retroviridae
Bovine leukosis virus	**see** Bovine leukaemia virus
Bovine oncovirus	**see** Bovine leukaemia virus
Breakbone fever virus	**see** Dengue virus
Brus–Laguna virus	*Bunyavirus*, Bunyaviridae
Bunyamwera virus	*Bunyavirus*, Bunyaviridae

C

Cache Valley virus	*Bunyavirus*, Bunyaviridae
California encephalitis virus	*Bunyavirus*, Bunyaviridae
Central European encephalitis virus	*Flavivirus*, Flaviviridae
Chikungunya virus	*Alphavirus*, Togaviridae
Colorado tick fever virus	– , Reoviridae
Congo virus	**see** Crimean–Congo haemorrhagic fever virus
Crimean–Congo haemor- rhagic fever virus	*Nairovirus*, Bunyaviridae

D

D'Aguilar virus	*Orbivirus*, Reoviridae
Dengue virus	*Flavivirus*, Flaviviridae
Dhori virus	– , (Orthomyxoviridae)
Durania virus	*Phlebovirus*, Bunyaviridae

E

Eastern equine encephalo- myelitis virus	*Alphavirus*, Togaviridae
EHD virus	**see** Epizootic haemorrhagic disease of deer virus
Encephalomyocarditis virus	*Cardiovirus*, Picornaviridae
Epidemic jaundice virus	**see** Hepatitis A virus
Epizootic haemorrhagic disease of deer virus	*Orbivirus*, Reoviridae

F

Far Eastern encephalitis virus	*Flavivirus*, Flaviviridae
Fort Sherman virus	*Bunyavirus*, Bunyaviridae

G

Gamboa virus	*Bunyavirus*, Bunyaviridae
Getah virus	*Alphavirus*, Togaviridae

H

Hantaan virus*	*Hantavirus*, Bunyaviridae
Hart Park virus	– , Rhabdoviridae
Hepatitis A virus	*Enterovirus*, Picornaviridae
Hepatitis B virus	– , Hepadnaviridae
Highlands J virus	*Alphavirus*, Togaviridae
Hundskrankheitvirus	**see** Sandfly fever virus

I

Igbo–Ora virus	*Alphavirus*, Togaviridae
Ilheus virus	*Flavivirus*, Flaviviridae
Infectious hepatitis virus	**see** Hepatitis A virus
Ixcanal virus	*Phlebovirus*, Bunyaviridae

J

Japanese encephalitis virus	*Flavivirus*, Flaviviridae

K

Karelian fever virus	*Alphavirus*, Togaviridae
Kemerovo virus	*Orbivirus*, Reoviridae
Kindia virus	*Orbivirus*, Reoviridae
Klamath virus	*Vesiculovirus*, Rhabdoviridae
Kunjin virus	*Flavivirus*, Flaviviridae
Kyasanur Forest disease virus	*Flavivirus*, Flaviviridae

L

La Crosse virus	*Bunyavirus*, Bunyaviridae
Langat virus	*Flavivirus*, Flaviviridae
Llano Seco virus	*Orbivirus*, Reoviridae
Louping ill virus	*Flavivirus*, Flaviviridae

M

Maguari virus	*Bunyavirus*, Bunyaviridae
Main Drain virus	*Bunyavirus*, Bunyaviridae
Malpais Spring virus	*Vesiculovirus*, Rhabdoviridae
Mayaro virus	*Alphavirus*, Togaviridae
Meaban virus	*Flavivirus*, Flaviviridae
Moellez virus	*Orbivirus*, Reoviridae
Murray Valley encephalitis virus	*Flavivirus*, Flaviviridae
Myxoma virus	*Leporipoxvirus*, Poxviridae

N

Negishi virus	*Flavivirus*, Flaviviridae
Northway virus	*Bunyavirus*, Bunyaviridae

O

Ockelbo virus	*Alphavirus*, Togaviridae
Omsk haemorrhagic fever virus	*Flavivirus*, Flaviviridae
O'nyong–nyong virus	*Alphavirus*, Togaviridae

P

Palyam virus	*Orbivirus*, Reoviridae
Powassan virus	*Flavivirus*, Flaviviridae

R

Rift Valley fever virus	*Phlebovirus*, Bunyaviridae
Ross River virus	*Alphavirus*, Togaviridae
Russian spring–summer enceph-alitis virus	**see** Far Eastern encephalitis virus

S

St. Louis encephalitis virus	*Flavivirus*, Flaviviridae
Sandfly fever Naples virus	*Phlebovirus*, Bunyaviridae
Sandfly fever Sicilian virus	*Phlebovirus*, Bunyaviridae
Semliki Forest virus	*Alphavirus*, Togaviridae
Serum hepatitis virus	**see** Hepatitis B virus
Sindbis virus	*Alphavirus*, Togaviridae
Snowshoe hare virus	*Bunyavirus*, Bunyaviridae
Soldado virus	*Nairovirus*, Bunyaviridae

T

Tahyna virus	*Bunyavirus*, Bunyaviridae
Tensaw virus	*Bunyavirus*, Bunyaviridae
Thimiri virus	*Bunyavirus*, Bunyaviridae
Thogoto virus	– , (Orthomyxoviridae)
Tickborne encephalitis virus	**see** Central European encephalitis virus
Trivittatus virus	*Bunyavirus*, Bunyaviridae
Turlock virus	*Bunyavirus*, Bunyaviridae

U

Uukuniemi virus	*Phlebovirus*, Bunyaviridae

V

Venezuelan equine encephalitis virus	*Alphavirus*, Togaviridae
Vesicular stomatitis virus	*Vesiculovirus*, Rhabdoviridae

W

Warthog virus	**see** African swine fever virus
Wesselsbron virus	*Flavivirus*, Flaviviridae
West Nile virus	*Flavivirus*, Flaviviridae
Western equine encephalitis virus	*Alphavirus*, Togaviridae

Y

Yellow fever virus	*Flavivirus*, Flaviviridae

Z

Zaliv Terpeniya virus	*Uukuvirus*, Bunyaviridae

*(not considered an arbovirus, although mites have been implicated in its transmission in China)

FUNGI

The following are the most important entomogenous fungi encountered in medical and veterinary entomology. The family and order/another taxon of each is given to the right, except where a synonym is present.

A

Aphanomyces   Saprolegniaceae, Saprolegniales
Aphanomyces astaci Schikora
Aphanomyces daphniae Prowse

Aspergillus   Moniliaceae, Hyphomycetales
Aspergillus flavus Link
Aspergillus fumigatus Fresenius
Aspergillus niger van Tieghem
Aspergillus parasiticus Speare
Aspergillus penicilloides Spegazzini
Aspergillus sydowii (Bainier & Sartory) Thom & Church
Aspergillus terreus Thom

Atkinsiella   Leptolegniellaceae, Saprolegniales
Atkinsiella entomophaga W.W. Martin

B

Beauveria   Moniliaceae, Hyphomycetales
Beauveria bassiana (Balsamo) Vuillemin
Beauveria brongniartii (Saccardo) Petch

C

Catenaria   Catenariaceae , Blastocladiales
Catenaria spinosa W.W. Martin

Coelomomyces   Blastocladiaceae, Blastocladiales
Coelomomyces africanus Walker
Coelomomyces dodgei Couch
Coelomomyces iliensis Dubitskiĭ, Dzerzhinskii & Danebekov
Coelomomyces indicus Iyengar
Coelomomyces lacunosus Couch & Sousa
Coelomomyces madagascaricus Couch & Grjebine
Coelomomyces psorophorae Couch
Coelomomyces punctatus Couch
Coelomomyces stegomyiae Keilin

Coelomycidium   ?Chytridiaceae, Chytridiales
Coelomycidium simulii Debaisieux

Conidiobolus   Entomophthoraceae, Entomophthorales
Conidiobolus coronatus (Costantin) Batko
Conidiobolus destruens (Weiser & Batko) Ben–Ze'ev
Conidiobolus incongruus Drechsler
Conidiobolus obscurus (Hall & Dunn) Remaudière & S. Keller

Cordyceps   Clavicipitaceae, Clavicipitales
Cordyceps dipterigena Berkeley & Broome

Culicinomyces   Moniliaceae, Hyphomycetales
Culicinomyces bisporalis Sigler, Frances & Panter
Culicinomyces clavisporus Couch, Romney & B. Rao
Culicinomyces parasiticus (Barron) Sigler

E

Empusa Entomophthoraceae, Entomophthorales
Empusa muscae Cohn **see** *Entomophthora muscae*
Empusa scatophagae (Giard) A. Jaczewski **see** *Entomophthora scatophagae*

Entomophaga Entomophthoraceae, Entomophthorales
Entomophaga aulicae (Reichardt) Humber
Entomophaga caroliniana (Thaxter) R.A. Samson, H.C. Evans & Latgé
Entomophaga tabanivora (J.F. Anderson & Magnarelli) Humber

Entomophthora Entomophthoraceae, Entomophthorales
Entomophthora aquatica J.F. Anderson & Anagnostakis **see** *Erynia aquatica*
Entomophthora aulicae (Reichardt) Humber **see** *Entomophaga aulicae*
Entomophthora caroliniana Thaxter **see** *Entomophaga caroliniana*
Entomophthora conica Nowakowski **see** *Erynia conica*
Entomophthora coronata (Costantin) Kevorkian **see** *Conidiobolus coronatus*
Entomophthora culicis (Braun) Fresenius
Entomophthora dipterigena Thaxter **see** *Erynia dipterigena*
Entomophthora fresenii (Nowakowski) M. Gustafsson
Entomophthora muscae (Cohn) Fresenius
Entomophthora planchoniana Cornu
Entomophthora scatophagae Giard
Entomophthora tabanivora J.F. Anderson & Magnarelli **see** *Entomophaga tabanivora*

Erynia Entomophthoraceae, Entomophthorales
Erynia aquatica (J.F. Anderson & Anagnostakis) Humber
Erynia caroliniana (Thaxter) Remaudière & Hennebert **see** *Entomophaga caroliniana*
Erynia conica (Nowakowski) Remaudière & Hennebert
Erynia curvispora (Nowakowski) Nowakowski
Erynia dipterigena (Thaxter) Remaudière & Hennebert
Erynia petchii (Ben-Ze'ev & Kenneth) Ben-Ze'ev & Kenneth
Erynia radicans (Brefeld) Humber, Ben-Ze'ev & Kenneth
Erynia variabilis (Thaxter) Remaudière & Hennebert

H

Hirsutella Moniliaceae, Hyphomycetales

L

Lagenidium Lagenidiaceae, Lagenidiales
Lagenidium giganteum Couch

Leptolegnia Saprolegniaceae, Saprolegniales
Leptolegnia chapmanii R.L. Seymour

M

Metarhizium Moniliaceae, Hyphomycetales
Metarhizium anisopliae (Metschnikoff) Sorokin
Metarhizium flavoviride W. Gams & Rozsypal

N

Neozygites Neozygitaceae, Entomophthorales
Neozygites fresenii (Nowakowski) Remaudière & S. Keller **see** *Entomophthora fresenii*

Nomuraea Moniliaceae, Hyphomycetales
Nomuraea atypicola (Yasuda) R.A. Samson
Nomuraea rileyi (Farlow) R.A. Samson

P

Paecilomyces Moniliaceae, Hyphomycetales
Paecilomyces farinosus (Holm) A.H.S. Brown & G. Smith

S

Smittium   Legeriomycetaceae, Harpellales
Smittium coloradense M.C. Williams & Lichtwardt
Smittium culisetae Lichtwardt
Smittium perforatum M.C. Williams & Lichtwardt
Smittium pusillum Manier & Coste
Smittium simulii Lichtwardt
Smittium tipulidarum M.C. Williams & Lichtwardt
Smittium typhellum Manier & Coste

T

Tolypocladium   Moniliaceae, Hyphomycetales
Tolypocladium cylindrosporum W. Gams
Tolypocladium extinguens R.A. Samson & Soares
Tolypocladium inflatum W. Gams
Tolypocladium parasiticum Barron

V

Verticillium   Moniliaceae, Hyphomycetales
Verticillium lecanii (A. Zimmermann) Viégas

Z

Zoophthora   **see** *Erynia*
Zoophthora radicans (Brefeld) Batko   **see** *Erynia radicans*

HELMINTHS

The following are the most common helminths encountered in relation to medical and veterinary arthropods. The family and order/another taxon of each is given to the right, except where a synonym is present.

A

Acanthocheilonema — Onchocercidae, Nematoda
Acanthocheilonema perstans (Manson) — **see** Mansonella perstans
Acanthocheilonema viteae (Krepkogorskaya) — **see** Dipetalonema viteae

Achillurbainia — Acillurbainiidae, Plagiorchiida
Achillurbainia congolensis (Fain & Vandepitte)

Agamomermis — Mermithidae, Nematoda
Agamomermis culicis Stiles — **see** Perutilimermis culicis

Anoplocephala — Anoplocephalidae, Cestoda
Anoplocephala magna (Abildgaard)
Anoplocephala perfoliata (Goeze)

Anoplocephaloides — Pomphorhynchidae, Echinorhynchida
Anaplocephaloides mamillana (Mehlis)
Anoplocephaloides ryjikovi (Spasskii)
Anoplocephaloides transversaria (Krabbe)

Aphanitylenchus — Parasitylenchidae, Nematoda

Avitellina — Anoplocephalidae, Cestoda
Avitellina arctica Kolmakov
Avitellina centripunctata (Rivolta)

B

Bertiella — Anoplocephalidae, Cestoda
Bertiella studeri (Blanchard)

Blatticola — Thelastomatidae, Nematoda
Blatticola barryi Zervos

Brugia — Onchocercidae, Nematoda
Brugia beaveri Ash & Little
Brugia malayi (Brug)
Brugia pahangi (Buckley & Edeson)
Brugia patei (Buckley et al.)
Brugia timori Partono et al.

C

Cardiofilaria — Onchocercidae, Nematoda
Cardiofilaria nilesi Dissanaike & Fernando

Cercopithofilaria — Onchocercidae, Nematoda
Cercopithofilaria roussilhoni Bain et al.

Choanotaenia — Dilepididae, Cestoda
Choanotaenia crateriformis (Goeze) — **see** Monopylidium crateriformis
Choanotaenia infundibulum (Bloch)
Choanotaenia sternina (Krabbe)

Cotugnia — Davaineidae, Cestoda
Cotugnia digonopora (Pasquale)

Culicimermis *Culicimermis schakhovii* Rubtsov & Issajeva	Mermithidae, Nematoda
Cyrnea *Cyrnea eurycerca* Seurat *Cyrnea mansioni* (Seurat)	Habronematidae, Nematoda

D

Dicrocoelium *Dicrocoelium dendriticum* (Rudolphi) *Dicrocoelium lanceolatum* (Rudolphi)	Dicrocoeliidae, Digenea **see** *Dicrocoelium lanceolatum*
Dipetalonema *Dipetalonema gracile* (Rudolphi) *Dipetalonema perstans* (Manson) *Dipetalonema reconditum* (Grassi) *Dipetalonema robini* Petit *et al.* *Dipetalonema streptocercus* (Macfie & Corson) *Dipetalonema viteae* (Krepkogorskaya)	Onchocercidae, Nematoda **see** *Mansonella perstans* **see** *Mansonella streptocerca*
Diphyllobothrium *Diphyllobothrium latum* (Linnaeus)	Diphyllobothriidae, Cestoda
Diplopylidium *Diplopylidium acanthotetra* (Parona)	Dilepididae, Cestoda
Dipylidium *Dipylidium caninum* (Linnaeus)	Dilepididae, Cestoda
Dirofilaria *Dirofilaria corynodes* (Linstow) *Dirofilaria immitis* (Leidy) *Dirofilaria magnilarvatum* Price *Dirofilaria repens* Railliet & Henry *Dirofilaria scapiceps* (Leidy) *Dirofilaria subdermata* (Monnig) *Dirofilaria tenuis* Chandler	Onchocercidae, Nematoda
Dracunculus *Dracunculus medinensis* (Linnaeus)	Dracunculidae, Nematoda

E

Elaeophora *Elaeophora schneideri* Wehr & Dikmans	Onchocercidae, Nematoda
Empidomermis *Empidomermis riouxi* Doucet *et al.*	Mermithidae, Nematoda
Enterobius *Enterobius vermicularis* (Linnaeus)	Oxyuridae, Nematoda
Eufilaria *Eufilaria kalifai* Millet & Bain	Onchocercidae, Nematoda
Eurytrema *Eurytrema coelomaticum* (Giard & Biller) *Eurytrema pancreaticum* (Janson)	Dicrocoeliidae, Digenea

F

Foleyella *Foleyella agamae* (Rodhain) *Foleyella candezei* (Fraipont)	Onchocercidae, Nematoda

G

Gastromermis
Gastromermis kolleonis Doucet & Poinar
Gastromermis mesostoma Poinar & Takaoka

Mermithidae, Nematoda

Gnathostoma
Gnathostoma hispidum Fedtschenko

Gnathostomatidae, Nematoda

Gongylonema
Gongylonema neoplasticum (Fibiger & Ditlevsen)
Gongylonema pulchrum Molin

Gongylonematidae, Nematoda

H

Habronema
Habronema muscae (Carter)

Habronematidae, Nematoda

Hammerschmidtiella
Hammerschmidtiella bareillyi Sharma & Gupta
Hammerschmidtiella diesingi (Hammerschmidt)

Thelastomatidae, Nematoda

Heleidomermis
Heleidomermis magnapapula Poinar & Mullens
Heleidomermis vivipara Rubtsov

Mermithidae, Nematoda

Heterorhabditis
Heterorhabditis heliothidis (Khan et al.)

Heterorhabditidae, Nematoda

Heterotylenchus
Heterotylenchus autumnalis Nickle

Sphaerulariidae, Nematoda

Hexamermis
Hexamermis glossinae Poinar et al.

Mermithidae, Nematoda

Howardula
Howardula prima Rubtsov
Howardula stenolobius Rubtsov

Allantonematidae, Nematoda

Hymenolepis
Hymenolepis diminuta (Rudolphi)
Hymenolepis nana (Siebold)

Hymenolepididae, Cestoda

see Vampirolepis nana

I

Isomermis
Isomermis bipapillatus Poinar & Takaoka
Isomermis lairdi Mondet et al.

Mermithidae, Nematoda

J

Joyeuxiella
Joyeuxiella echinorhynchoides (Sonsino)
Joyeuxiella pasqualei (Diamare)

Dilepididae, Cestoda

L

Leidynema
Leidynema appendiculatum (Leidy)

Thelastomatidae, Nematoda

Ligula
Ligula intestinalis (Linnaeus)

Diphyllobothriidae, Cestoda

Litomosoides
Litomosoides carinii (Travassos)

Onchocercidae, Nematoda

Loa
Loa loa (Cobbold)

Onchocercidae, Nematoda

M

Macracanthorhynchus
Macracanthorhynchus hirudinaceus (Pallas)

Oligacanthorhynchidae, Acanthocephala

Mansonella
Mansonella colombiensis (Esslinger)
Mansonella mariae Petit *et al.*
Mansonella ozzardi (Manson)
Mansonella perstans (Manson)
Mansonella streptocerca (Macfie & Corson)

Onchocercidae, Nematoda

Mesocestoides
Mesocestoides lineatus (Goeze)

Mesocestoididae, Cestoda

Mermis
Mermis nigrescens Dujardin

Mermithidae, Nematoda

Mesomermis
Mesomermis crassivaginae Camino
Mesomermis dissimilis Camino
Mesomermis ochrae Camino
Mesomermis odeschti Gafurov
Mesomermis subandina Camino

Mermithidae, Nematoda

Molinema
Molinema arbuta (Highby)

Onchocercidae, Nematoda

Monanema
Monanema marmotae (Webster)
Monanema martini Bain *et al.*
Monanema nilotica El-Bihari *et al.*

Onchocercidae, Nematoda

Moniezia
Moniezia autumnalia Kuznetsov
Moniezia benedeni (Moniez)
Moniezia expansa (Rudolphi)

Anoplocephalidae, Cestoda

Moniliformis
Moniliformis dubius Meyer
Moniliformis moniliformis (Bremser)

Moniliformidae, Acanthocephala
see *Moniliformis moniliformis*

N

Neoaplectana
Neoaplectana bibionis (Bovien)
Neoaplectana carpocapsae (Weiser)
Neoaplectana feltiae (Filipjev)
Neoaplectana glaseri (Steiner)
Neoaplectana intermedia Poinar

Steinernematidae, Nematoda
see *Neoaplectana feltiae*

O

Octomyomermis
Octomyomermis albicans Camino
Octomyomermis muspratti (Obiamiwe & MacDonald)
Octomyomermis troglodytis Poinar & Sanders

Mermithidae, Nematoda

see *Romanomermis muspratti*
see *Romanomermis troglodytis*

Onchocerca
Onchocerca armillata Railliet & Henry
Onchocerca cervicalis Railliet & Henry
Onchocerca gibsoni (Cleland & Johnson)
Onchocerca gutturosa Neumann
Onchocerca lienalis (Stiles)
Onchocerca ochengi Bwangamoi
Onchocerca tarsicola Bain & Schulz-Key

Onchocercidae, Nematoda

Onchocerca volvulus (Leuckart)

Oxyspirura   Thelaziidae, Nematoda
Oxyspirura mansoni (Cobbold)

P

Panagrolaimus   Panagrolaimidae, Nematoda
Panagrolaimus migophilus Poinar & Geetha Bai

Paragonimus   Paragonimidae, Digenea
Paragonimus africanus Voelker & Vogel
Paragonimus amazonicus Miyazaki *et al.*
Paragonimus caliensis Little
Paragonimus congolensis (Fain & Vandepitte)   **see** *Achillurbainia congolensis*
Paragonimus ecuadoriensis Voelker & Arzube
Paragonimus heterotremus Chen & Hsia
Paragonimus hueitungensis Chung *et al.*
Paragonimus kellicotti Ward
Paragonimus mexicanus Miyazaki & Ishii
Paragonimus miyazakii Kamo *et al.*
Paragonimus peruvianus Miyazaki *et al.*   **see** *Paragonimus mexicanus*
Paragonimus philippinensis Ito *et al.*   **see** *Paragonimus westermani*
Paragonimus pulmonalis (Baelz)
Paragonimus ringeri Cobbold   **see** *Paragonimus westermani*
Paragonimus skrjabini Chen
Paragonimus uterobilateralis Voelker & Vogel
Paragonimus westermani (Kerbert)

Paranoplocephala   Anoplocephalidae, Cestoda
Paranoplocephala mamillana (Mehlis)   **see** *Anoplocephaloides mamillana*
Paranoplocephala ryjikovi Spasskii   **see** *Anoplocephaloides ryjikovi*
Paranoplocephala transversaria (Krabbe)   **see** *Anoplocephaloides transversaria*

Paricterotaenia   Dilepididae, Cestoda
Paricterotaenia sternina (Krabbe)   **see** *Choanotaenia sternina*

Pelecitus   Onchocercidae, Nematoda
Pelecitus fulicaeatrae (Diesing)

Perutilimermis   Mermithidae, Nematoda
Perutilimermis culicis (Stiles)

Pheromermis   Mermithidae, Nematoda
Pheromermis montanus Gafurov & Muratova
Pheromermis robustus Gafurov & Muratova
Pheromermis rubzovi Andreeva & Spiridonov
Pheromermis tabanivora Rubtsov & Andreeva

Plagiorchis   Plagiorchiidae, Digenea
Plagiorchis noblei Park

Pleurogenoides   Lecithodendriidae, Digenea
Pleurogenoides orientalis Srivastava

Polymorphus   Centrorhynchidae, Acanthocephala
Polymorphus contortus (Bremser)

Pomphorhynchus   Pomphorhynchidae, Acanthocephala
Pomphorhynchus bulbocolli Linkins

Protrellus   Thelastomatidae, Nematoda
Protrellus dalei Zervos
Protrellus dixoni Zervos

Psyllotylenchus
Psyllotylenchus chabaudi Deunff & Launay

Allantonematidae, Nematoda

R
Raillietina
Raillietina cesticillus (Molin)
Raillietina echinobothrida (Megnin)
Raillietina tetragona (Molin)

Davaineidae, Cestoda

Raphidia
Raphidia notata Fabricius
Raphidia xanthostigma Schummel

Raphidiidae, Raphidioptera

Rhabditis

Rhabditidae, Nematoda

Romanomermis
Romanomermis altaica Gubaidulin & Vakker
Romanomermis communensis Galloway & Brust
Romanomermis culicivorax (Ross & Smith)
Romanomermis iyengari Welch
Romanomermis jingdeensis Yang & Chen
Romanomermis muspratti (Obiamiwe & MacDonald)
Romanomermis nielseni (Tsai & Grundmann)
Romanomermis sichuanensis Peng & Song
Romanomermis troglodytis (Poinar & Sanders)
Romanomermis yunanensis Song & Peng

Mermithidae, Nematoda

S
Schistocephalus
Schistocephalus pungitii Dubinina
Schistocephalus solidus (Müller)

Diphyllobothriidae, Cestoda

Schwenkiella
Schwenkiella atheri Parveen & Jairajpuri

Thelastomatidae, Nematoda

Setaria
Setaria labiatopapillosa (Alessandrini)

Onchocercidae, Nematoda

Spilotylenchus
Spilotylenchus beaucournui Launay & Deunff

Allantonematidae, Nematoda

Spirometra
Spirometra mansoni (Cobbold)
Spirometra mansonoides (Müller)

Diphyllobothriidae, Cestoda

Spirura
Spirura gastrophila Müller
Spirura rytipoeurites (Delongchamps)
Spirura talpae (Gmelin)

Spiruridae, Nematoda
see *Spirura talpae*
see *Spirura talpae*

Steinernema
Steinernema bibionis (Bovien)
Steinernema carpocapsae Weiser
Steinernema feltiae (Filipjev)
Steinernema glaseri (Steiner)

Steinernematidae, Nematoda
see *Neoaplectana feltiae*
see *Neoaplectana carpocapsae*
see *Neoaplectana feltiae*
see *Neoaplectana glaseri*

Stephanofilaria
Stephanofilaria assamensis Pande
Stephanofilaria kaeli Buckley
Stephanofilaria stilesi Chitwood

Filariidae, Nematoda

Stilesia
Stilesia globipunctata (Rivolta)
Stilesia hepatica Wolffhügel

Anoplocephalidae, Cestoda

Strelkovimermis	Mermithidae, Nematoda
Strelkovimermis peterseni (Nickle)	
Strelkovimermis singularis (Strelkov)	
Strelkovimermis spiculatus Poinar & Camino	
Suifunema	Thelastomatidae, Nematoda
Suifunema mackenziei Zervos	

T

Taenia	Taeniidae, Cestoda
Taenia hydatigena Pallas	
Taenia pisiformis (Bloch)	
Taenia solium Linnaeus	
Tetradonema	Tetradonematidae, Nematoda
Tetradonema solenopsis Nickle & Jouvenaz	
Tetrameres	Tetrameridae, Nematoda
Tetrameres inermis (Linstow)	
Tetrapetalonema	Onchocercidae, Nematoda
Tetrapetalonema perstans (Manson)	**see** *Mansonella perstans*
Tetrapetalonema streptocerca (Macfie & Corson)	**see** *Mansonella streptocercus*
Thelastoma	Thelastomatidae, Nematoda
Thelastoma bulhoesi (Magalhães)	
Thelazia	Thelaziidae, Nematoda
Thelazia callipaeda Railliet & Henry	
Thelazia gulosa Railliet & Henry	
Thelazia rhodesii Demarest	
Thelazia skrjabini Erschoff	
Thysaniezia	Anoplocephalidae, Cestoda
Thysaniezia giardi (Moniez)	**see** *Thysaniezia ovilla*
Thysaniezia ovilla (Rivolta)	
Thysanosoma	Anoplocephalidae, Cestoda
Thysanosoma actinioides (Diesing)	
Triaenophorus	Triaenophoridae, Cestoda
Triaenophorus crassus Forel	
Triaenophorus lucii (Mueller)	
Triaenophorus nodulosus (Pallas)	**see** *Triaenophorus lucii*
Trichocephalus	Trichuridae, Nematoda
Trichocephalus trichiurus (Linnaeus)	**see** *Trichuris trichiura*
Trichuris	Trichuridae, Nematoda
Trichuris trichiura (Linnaeus)	

V

Vampirolepis nana	Hymenolepididae, Cestoda

W

Wuchereria	Onchocercidae, Nematoda
Wuchereria bancrofti (Cobbold)	
Wuchereria kalimantani Palmieri *et al.*	
Wuchereria malayi (Brug)	**see** *Brugia malayi*

Y

Yatesia	Onchocercidae, Nematoda
Yatesia hydrochoerus (Yates)	

FISH

The following are the most important fish used in medical and veterinary entomology for biological control. The family and order/another taxon of each is given to the right, except where a synonym is present.

A
Aphanius Cyprinodontidae, Cyprinodontiformes
Aphanius dispar (Rüppell)
Aphanius iberus (Cuvier & Valenciennes)

Aphyocypris Cyprinidae, Cypriniformes
Aphyocypris chinensis Günther

Aplocheilus Aplocheilidae, Cyprinodontiformes
Aplocheilus latipes Temminck & Schlegel **see** *Oryzias latipes*
Aplocheilus panchax (Hamilton-Buchanan)

C
Carassius Cyprinidae, Cypriniformes
Carassius auratus Linnaeus

Cyprinodon Cyprinodontidae, Cyprinodontiformes
Cyprinodon dispar Rüppell **see** *Aphanius dispar*
Cyprinodon iberus Cuvier & Valenciennes **see** *Aphanius iberus*

E
Esomus Cyprinidae, Cypriniformes
Esomus danricus (Hamilton-Buchanan)

G
Gambusia Poeciliidae, Cyprinodontiformes
Gambusia aculeatus
Gambusia affinis Baird & Girard
Gambusia affinis holbrooki Girard **see** *Gambusia holbrooki*
Gambusia holbrooki Girard
Gambusia punctata Poey
Gambusia puncticulata Poey

H
Haplochilus Aplocheilidae, Cyprinodontiformes
Haplochilus panchax (Hamilton-Buchanan) **see** *Aplocheilus panchax*

Haplochromis Cichlidae, Perciformes

L
Lebistes Poeciliidae, Cyprinodontiformes
Lebistes reticulatus (Peters) **see** *Poecilia reticulata*

Lucania Fundulidae, Cyprinodontiformes
Lucania parva Baird & Girard

N
Nothobranchius Aplocheilidae, Cyprinodontiformes
Nothobranchius korthausae Meinken

Nuria Cyprinidae, Cypriniformes
Nuria danrica Hamilton-Buchanan **see** *Esomus danricus*

O

Oreochromis
Oreochromis aureus (Steindachner)
Oreochromis mossambicus (Peters)
Oreochromis niloticus (Linnaeus)
Oreochromis spilurus (Günther)

Cichlidae, Perciformes

Oryzias
Oryzias javanicus (Bleeker)
Oryzias latipes (Temminck & Schlegel)

Oryziatidae, Cyprinodontiformes

P

Pachypanchax
Pachypanchax playfairi (Günther)

Aplocheilidae, Cyprinodontiformes

Panchax
Panchax panchax Hamilton-Buchanan
Panchax playfairi (Günther)

Cyprinodontidae, Cyprinodontiformes
see *Aplocheilus panchax*
see *Pachypanchax playfairi*

Parluciosoma
Parluciosoma daniconius (Hamilton-Buchanan)

Cyprinidae, Cypriniformes

Pimephales
Pimephales promelas Rafinesque

Cyprinidae, Cypriniformes

Poecilia
Poecilia reticulata (Peters)

Poeciliidae, Cyprinodontiformes

Pseudomugil
Pseudomugil signifer Kner

Atherinidae, Perciformes

R

Rasbora
Rasbora daniconius (Hamilton-Buchanan)

Cyprinidae, Cypriniformes
see *Parluciosoma daniconius*

Rivulus
Rivulus marmoratus Poey
Rivulus milesi Fowler

Rivulidae, Cyprinodontiformes

S

Sarotherodon

see *Oreochromis*

T

Tilapia
Tilapia aurea Steindachner
Tilapia mossambica Peters
Tilapia nilotica Linnaeus
Tilapia spiluris Günther

see *Oreochromis*
see *Oreochromis aureus*
see *Oreochromis mossambicus*
see *Oreochromis niloticus*
see *Oreochromis spilurus*

X

Xenotoca
Xenotoca eiseni (Rutter)

Goodeidae, Cyprinodontiformes

OTHER ORGANISMS

The following are other organisms encountered in medical and veterinary entomology, either as parasites, predators or pathogens. The family and order/another taxon of each is given to the right, except where a synonym is present.

D
Dugesia   Dugesiidae, Tricladida
Dugesia dorotocephala (Woodworth)
Dugesia japonica Ichikawa & Kawakatsu
Dugesia tigrina (Girard)

L
Lambornella   Tetrahymenidae, Ciliophora
Lambornella clarki Corliss & Coats

M
Mesostoma   Typhloplanidae, Tricladida
Mesostoma californicum Hyman
Mesostoma lingua (Abildgaard)
Mesostoma vernale Hyman
Mesostoma zariae Kolasa & Mead

N
Nephelopsis   Erpobdellidae, Hirudinea
Nephelopsis obscura Verrill

T
Temnocephala   Temnocephalidae, Tricladida
Temnocephala chaeropsis